Matter Matters?

Springer-Verlag Berlin Heidelberg GmbH

Peter Århem
Hans Liljenström
Uno Svedin (Eds.)

Matter Matters?

On the Material Basis of the Cognitive Activity of Mind

With 67 Figures

Springer

Editors:

Peter Århem
The Nobel Institute for Neurophysiology
Karolinska Institutet
S-171 77 Stockholm, Sweden

Hans Liljenström
ANS – Studies of Artificial Neural Systems
Department of Numerical Analysis and Computing Science
Royal Institute of Technology
S-100 44 Stockholm, Sweden

Uno Svedin
Swedish Council for Planning and Coordination of Research (FRN)
P.O. Box 7101
S-103 87 Stockholm, Sweden

Technical Editor:

Mats-Olov Olsson
Centre for Regional Science (CERUM)
Umeå University
S-901 87 Umeå, Sweden

Library of Congress Cataloging-in-Publication Data

Matter matters? : on the material basis of the cognitive activity of mind / edited by Peter Århem, Hans Liljenström, Uno Svedin. p. cm.
ISBN 978-3-642-64459-7 (hardcover : alk. paper)
1. Science–Philosophy. 2. Mind and body. 3. Brain. I. Århem, Peter, 1946– . II. Liljenström, Hans, 1956– . III. Svedin, Uno, 1943– . Q175.M398 1997 153–dc21 96-52744 CIP

ISBN 978-3-642-64459-7 ISBN 978-3-642-60570-3 (eBook)
DOI 10.1007/978-3-642-60570-3

Softcover reprint of the hardcover 1st edition 1997

Typesetting: Data conversion by Kurt Mattes, Heidelberg
Cover design: Erich Kirchner, Heidelberg
SPIN 10425935 55/3144 – 5 4 3 2 1 0 – Printed on acid-free paper

Preface

Since 1983 the Swedish Council for Planning and Coordination of Research (FRN) has invited international scholars for workshops at the field station of the Royal Academy at Abisko regarding important crosscutting themes in science, often related to complexity topics. The remote and serene location above the polar circle and facing the mountain range has always contributed to their informal character and to lively discussions.

In 1993, the topic chosen was "Matter Matters? On the Material Basis of the Cognitive Activity of Mind". The idea was to dwell on the "eternal issue" of mind-brain and do it from a broad range of perspectives. One specific entry point, though, was to see to what extent recent advances in mathematics, physics, and biosciences can contribute to a deeper understanding of the entire complex.

The organizing committee was Prof. Uno Svedin (chair), Ass. Prof. Hans Liljenström, Mr. Mats-Olov Olsson (conference manager), Dr. Rolf Wasén, and Ass. Prof. Peter Århem. In the follow-up to the conference the organizing committee engaged in the production of this book. Special thanks go to Mr. Mats-Olov Olsson, who has served as technical editor.

Stockholm, March 1997

Uno Svedin
Hans Liljenström
Peter Århem

Contents

List of Authors

Peter Århem
The Nobel Institute
for Neurophysiology
Karolinska Institutet
S-171 77 Stockholm, Sweden

Gregory J. Chaitin
IBM Research Division
P. O. Box 704
Yorktown Heights, NY 10598
USA

Michael Conrad
Department of Computer Science
Wayne State University
Detroit, Michigan 48202, USA

Deborah Conrad
Department of Philosophy
Wayne State University
Detroit, Michigan 48202, USA

Hermann Haken
Institute for Theoretical Physics
and Synergetics
University of Stuttgart
Pfaffenwaldring 57/4
D-70550 Stuttgart, Germany

Stuart R. Hameroff
Departments of Anesthesiology
and Psychology
University of Arizona
Health Sciences Center
Tucson, Arizona 85724, USA

Hans Liljenström
SANS – Studies of Artificial Neural Systems
Dept. of Numerical Analysis
and Computing Science
Royal Institute of Technology
S-100 44 Stockholm, Sweden

B. I. B. Lindahl
Department of Philosophy
University of Stockholm
S-106 91 Stockholm, Sweden

Roger Penrose
Mathematical Institute
University of Oxford
24-29 St. Giles
Oxford OX1 3LB, England

Robert Rosen
Department of Physiology and Biophysics
Faculty of Medicine
Dalhousie University
Halifax, Nova Scotia B3H 4H7, Canada

Petra Stoerig
Institute of Medical Psychology
Ludwig-Maximilians-University
Goethestr. 31
D-80336 Munich, Germany

Uno Svedin
Swedish Council for Planning
and Coordination of Research (FRN)
P. O. Box 7101
S-103 87 Stockholm, Sweden

1. Introduction

Peter Århem, Hans Liljenström, and Uno Svedin

The phenomenon of mind and its connection to the brain has never ceased to be of great concern to mankind. However, different historical epochs have had their own capacity – or lack of capacities – to address this issue. The cultural framework, which includes the state of the art of science of today, as well as its associated world views, has also heavily framed the discourse.

Thus, it is no surprise that also in our time this eternal issue should appear with its own specific characteristics. What is special for our century is the new capacity to successfully address issues at the microscopic level of matter. The second part of this century has also seen an entirely new understanding emerge, concerning fundamental issues about computability and related questions. Various experimental techniques and theoretical developments have further allowed a deeper understanding of the great complexity of the mammalian brain. The area of investigation surrounding the mind-matter issue has thus during this century been provided with entirely new and fresh insights from domains that earlier were not developed.

This book tries to encompass several of these new domains of knowledge in an attempt to provide the reader with a systematic representation of perspectives relevant for an understanding of the mind-matter issue. It would be wrong to say that there is a scientific consensus about these issues. In fact, this is also a reason for a collection of texts that provide insights from many disciplines and at different levels of approaches to these problems.

The phrase "matter matters", in the context of the mind-matter issue, indicates a focal interest on the material side and its importance for understanding mind processes. However, matter appears at different systemic levels, i. e., the way matter is organized has importance. Thus, the approach involving a certain emphasis on the material side in the mind-matter dichotomy is complemented by another approach, namely via the organizational, structural, and systemic issues. Matter matters.

Yes, but it does so through the material form constraints it poses on the organizational architecture of the brain. And the systems features may offer clues to the understanding of cognition phenomena that are as interesting as either material component approaches or those with direct focus on mental structures or components or dynamics.

The book is organized into four blocks. The first one addresses some general principles at the foundation of the mind-matter issue from the points of view of physics, computer science, and theoretical biology. In so doing it introduces some new perspectives that shift the focus of interest from more traditional ways of viewing the problems to other, perhaps more fertile ways.

Penrose uses a physics approach in an attempt to relate awareness and physical computational issues in the scientific realm, shifting the focus from the classical dichotomy material-immaterial to that of computable versus non-computable. Chaitin, one of today's successors to Gödel, shifts the focus from more traditional ways of treating the well-known Gödel theorem to other aspects by introducing the Berry paradox. Rosen, from his biological standpoint, introduces new ways of viewing phenotypic and genotypic processes.

Following the perspectives given in the first block, where the mind-matter problem is treated in a theoretical and more general way, the following blocks are more specific and direct. The contributions in these blocks deal with the problem at different organizational levels, from the microscopic to the macroscopic.

The sub-cellular level is treated in the second block. Hameroff discusses effects of quantum coherence in the cytoskeleton as crucial for various aspects of consciousness. Conrad & Conrad in their article also point to the specific and unique physical-chemical-dynamical nature of living organisms as essential, and inimitable by any artificial system.

In the third block the network and systems level is treated. Haken applies synergetics, his theory for self-organizing systems, to pattern recognition in the brain's visual system. He demonstrates how a computer system based on this theory can recognize human faces. Liljenström also uses computer models to test ideas of how the complex dynamics of the brain could have evolved for efficient information processing.

The fourth and final block treats the subjective level. Stoerig addresses the questions of how self and consciousness have evolved in parallel and which parts of the brain are involved in conscious visual perception in particular. Århem & Lindahl end the book with an article on the role of spontaneous neural activity for thinking and consciousness, giving supportive evidence for molecular events that could trigger neuronal impulses.

More detailed introductory guidance to these blocks will be given at the beginning of each one of them. In these guiding and linking segments of text we will also go back to the earlier parts of the book and see what we have learnt accumulatively. In this way, the book contains both the four blocks with their individual chapters written by individual authors, and also these connecting and commenting segments. A reader who wants a quick overview before reading every chapter could read the segment texts in sequence to get a first glance at the total content.

Before we enter the first block, i. e., the one about generality and basic perspectives, some common themes should be highlighted already at the beginning. Many of the key topics will appear over and over again. Thus, it is of interest to outline some of them here.

1.1 The Interplay Between Different Disciplines

As has already become clear, the aim of this book is to provide a range of different perspectives on a common theme. We have decided not to give hegemony to any academic discipline in the interpretation of the mind-matter problem. The reason is that a multiplicity of perspectives is necessary for such a complex issue. The various levels of interpretation are represented in different ways by different disciplines, but also their specific focus of interest varies. Already the interplay between physics, mathematics/computer science, and biology as prerequisites for the "matter entry point" provides ample examples for these often complementary views. The "mind entry point" could have been made more elaborate and diversified in itself, if we had not from the beginning chosen to stress that matter matters. Still, the special characteristics of the "mind side" should be recognized not to be reducible to those categories used in the brain-matter part of the spectrum.

1.2 Micro and Macro Levels

Different approaches choose to stress the different levels of aggregation, from a very atomistic micro-level to a macro-level of phenomena. In our understanding, each level has to be given a certain freedom of expression for the concepts specific to that level. This is done here by combining similar "level" approaches in the blocks mentioned above. However, we want to address equally well the interlinking between levels of interpretation, including feedbacks and feedforwards between these levels. Some of these aspects are dealt with by Haken and Liljenström in the block on the network and systems level.

Also, Penrose addresses these issues and calls for new theory elements in the meso-interval between the quantum-mechanical micro-world and that of more pronounced macro-phenomena. In this context, the issue of the quantum-theoretical implications for the entire scheme of interpretations of mind-brain phenomena is put on the table. The issue is basically treated by Penrose, but is also commented upon by Århem & Lindahl.

In discussing various spatial scales, it is equally important to consider the different time scales involved. In general, most processes and events at the microscopic level of molecules and ions occur at a very small time scale, within nanoseconds or less. Processes at more macroscopic levels, on the other hand, usually occur at larger time scales, milliseconds or longer. This topic is also treated in some of the contributions.

1.3 The Role of Mathematics

As must have become evident already, the framework of many of the considerations brought to bear on the topic of this book is mathematics. This holds true especially in those parts where parallels are drawn to computer modeling, both with regard to a possible better understanding of the "machinery" of the brain and its functions and also at a deeper level concerned with possibilities of understanding.

The relationship between the world of mathematics and that of the "real" world is not self-evident or straightforward. Many of the remarks made in the first block on generalities reflect on this issue, which could be labeled "the potential role of mathematics in understanding issues about brain-mind phenomena".

1.4 Complexity

This book tries to cut through many layers of interpretation. This is because the topic itself exposes many features of what could be called "complex" phenomena. Later in the book, in the specific contexts where this is discussed, a more precise definition is given. At this stage, though, it seems to be relevant to highlight that the topic of complexity is one of the common themes of the book, explicitly and implicitly.

In some of the articles it is pointed out that the matter of complexity is related to evolution, where for example the nervous system grows more complex in the course of evolution, allowing for more complex and elaborate behavior of the organism. The question of evolution is a common theme in many of the contributions, although from different points of view.

1.5 What Is Meant by Understanding?

In the further elaboration of the variety of approaches it will be evident that the understanding of mind-brain phenomena has many facets. But at a deeper level we talk about not only a multiplicity of approaches to an emerging understanding but also a challenge regarding different styles of understanding. What does it really mean that we have understood the relationship between the brain and the mind? Paradoxes may appear along the way as the expression of incompletely resolved schemes of interpretation, all different and each partially compelling. We hope this book will also provide elements for thought in this direction.

Part I

Some New Perspectives

In this first part, some aspects of the mind-matter issue will be considered from the points of view of mathematics, physics, computer science, and theoretical biology. The discourse will result in some new perspectives that will form the background to the more concrete contributions in the following blocks.

The physics/mathematics approach by Roger Penrose tries to relate awareness and physical computational issues in the scientific realm. He puts some hard questions about scientific explanation in general, and specifically, to what extent issues about computational explanations can be used to throw light on the mind-matter issue. This is important not only in view of the general scientific challenge of using mathematical tools to address important scientific issues but also in order to investigate the more specific issue of whether brain activities can be treated in terms of computer science methods. The issue of what mathematical models can tell us, and specifically what computability or non-computability means for the mind-brain debate, is highlighted.

A key issue here is the nature of human thinking. Can thinking generally be described as algorithmic, i. e., does it proceed in terms of computational processes or are non-computational components involved? The conventional view is reflected in the statement often made that all (biological) processes are algorithmic, thinking as well as evolution. This forms the basis for the dominant computer approach to the problem, irrespective of whether it is the classical Artificial Intelligence or the (now more fashionable) connectionist approach. This should be seen in sharp contrast to the view here presented by Penrose that conscious thinking, and other processes as well, involves principally non-computational components. Penrose bases his argument on mathematical thinking, probably the most analyzable and therefore the most rewarding area in this context. Others have based their arguments on other forms of reasoning (see Hodgson 1991).

The discussion on the algorithmic nature of mathematical reasoning naturally raises the related, but not identical, question whether brain processes in general can be fully described by algorithms or not. Assuming that conscious thinking is non-computational, but materially based (i. e., rooted in the brain), brain processes must, in principle, comprise at least some non-computational components. It is interesting to note that this question seems independent of the classical question of psycho-physical interactionism or parallelism as a solution to the mind-brain problem. Both these positions are compatible with the argument for brain processes of a non-computational kind. This insight seems essential. The mind-brain debate seems by this change in perspective to have reached a new, more fertile level than earlier, a level where the classical conflict between psycho-physical interactionism and parallelism has become obsolete and replaced by the more concrete debate about the possible degree of non-computability of mental processes.

The discussion by Penrose – relevant on its own ground – is also an important preparation for the account by Chaitin about central issues related to computability, here presented in a personalised style of discourse. Here, some words about the Gödel theorem are perhaps needed for the non-specialist reader. What Gödel did was to show that certain logical systems can never be said to be complete, despite all efforts we may put into such an endeavor. There are always questions that fall outside the realm of what can be answered within a certain set of axioms. Of course, when you find such questions you can expand the number of axioms in order to encompass the new questions, but only to find that still others fall outside what can be treated. It is a never-ending race to find a complete system that encompasses everything. Indeed, there is no such set to be found. This deep result of Gödel's was a fundamental contribution to mathematics and logic, but also to philosophy. This is the realm of computability raised by Penrose as a key issue in understanding the issue of consciousness in particular, and of the mind-matter issue in general.

In his chapter Chaitin discusses the development from Gödel and onwards. He introduces in the discourse about "limits of mathematics" an alternative to Gödel's use of the "paradox of the liar". Instead, Chaitin introduces the "Berry paradox". What is here being dwelt upon? The discussion is about what is meant by the term "false" and what it implies. Chaitin formulates it as something that "does not correspond to reality". Then, he moves in several steps to issues about what is "unprovable" and what is "incomplete". That, in turn, is just at the heart of Penrose's discussion about the role of mathematics in understanding mind-matter issues.

Rosen closes this first block on general principles with a discussion of the mind-brain issue as seen from a biological standpoint. Key issues are how reduction is achieved in scientific explanations, but also how different forms of aspired objectivity in various scientific approaches influence the issues. His reference to the Turing test, which is part of the discourse on computability opened by Gödel, could be seen as an entry point to the issue of what is meant by "intelligent behavior". However, at a deeper level it also connects to the issues earlier raised about mathematics as a framework for understanding key scientific issues, and computability and incompleteness issues in general. Rosen seems to go further than Penrose in his criticism of the mainstream view on algorithmic processes. By using the concepts 'phenotypic' and 'genotypic' processes, in a highly original way he concludes that algorithms can, in principle, fully describe neither mind nor biological processes. The reason is that algorithms inherently imply that the genotype and phenotype coincide and, in Rosen's words, we find ourselves in a world in which life and consciousness cannot be accommodated – indeed a very unusual and "simple" world.

The discussion of mathematical thinking in this block touches upon another key issue of this book, one at the foundation of the mind-matter problem, namely the nature of mathematical knowledge. Penrose seems to suggest the existence of an autonomous mathematical world, closely related to the world of mathematical forms of Plato (see Penrose 1994). A much discussed version of such a view was suggested by Popper (1972). He argued that there is an autonomous world of objective knowledge, created by man. In comparison with Popper's world, Penrose's mathematical world is closer to Plato's original world, in being eternal and not man-made. Mathematical theorems are always discoveries for Penrose, not inventions. For Popper, they are a combination of inventions and discoveries (Popper 1994).

Assuming that this, among philosophers somewhat unfashionable, view holds, we are led to the question (also stated by Einstein and others), "How can this autonomous mathematical world be so close to the physical world?" Or in other terms: why is it so that mathematics describes the physical world so well? This is even more enigmatic, when one considers that (as pointed out by Chaitin) mathematics is much simpler than the world.

This remarkable connection between mathematics and reality is highlighted in Penrose's attempt to find non-computable physical processes. In this, he adopts a quantum-mechanical approach, arguing that quantum mechanics contains non-deterministic features in the reduction process associated with measurements, i. e., in terms of the collapse of the wave function. But he seems to go further, as he also suggests that certain microstructures of the nerve cells are the physical correlates to such non-computational processes. By using arguments from an abstract mathematical level as well as from a concrete biological level, Penrose highlights one of the themes of this book, the necessity of a multilevel approach to the mind-matter problem. This discussion also forms a natural bridge to the next block in this book, that on sub-cellular issues.

References

Hodgson, D. (1991). *The Mind Matters*, Clarendon Press, Oxford

Penrose, R. (1994). *Shadows of the Mind*, Oxford University Press, Oxford

Popper, K. R. (1972). *Objective Knowledge: An Evolutionary Approach*, Clarendon Press, Oxford

Popper, K. R. (1994). *Knowledge and the Body-Mind Problem: In Defense of Interaction*, Routledge, London

2. The Need for a Non-computational Extension of Quantum Action in the Brain

Roger Penrose

2.1 Introduction

In this chapter I want to discuss a question that has to do with whether quantum theory and quantum computers would in some kind of miraculous way help us understand the mind-brain problem. As far as I understand Dennett's usage, it would *not* be a skyhook, but I shall give the reasons for this position and later on explain what I mean by quantum processes being relevant to whatever it is that might be going on in the brain.

There is one point that should be stressed here, which often has been misunderstood. The issue that I am really concerned with in the *Emperor's New Mind* (Penrose 1989), and even more so in *Shadows of the Mind* (Penrose 1994), has to do with the difference between understanding something scientifically or mathematically, and having a computational explanation of something.

I want to stress that computation is a very particular area of mathematics. In the large space of everything that belongs to mathematics, there are certain things that are based on computation, but that would only be some tiny little point in that large space (see Fig. 2.1). Even if you restrict yourself to ordinary numbers, those which are of computational character would be a tiny limited set. I shall make that more explicit later.

I think that because computers are so important in present-day science, people almost identify scientific explanation with computational explanation. Although it is true that most of what is done in present-day science is illuminated to a great degree by computers – they are of great importance to what is going on nowadays – we have to realise that a computational explanation is a quite different thing from a scientific explanation, or from a mathematical explanation. I think it is important to make that distinction.

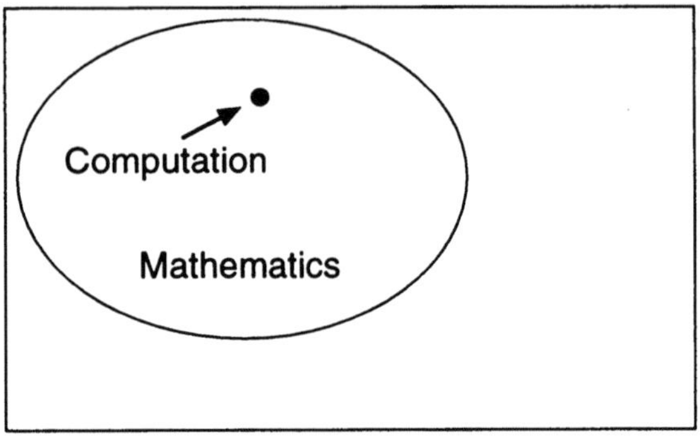

Fig. 2.1. Computation is just a tiny part of mathematics

2.2 Views on Consciousness

We now examine some of the major viewpoints concerning the relationship between consciousness and physical computational processes. Four alternative viewpoints are listed below:

A) All thinking is computation and the mere carrying out of appropriate computations will evoke feelings of conscious awareness.
B) It is the brain's physical action that evokes awareness and any physical action can in principle be simulated computationally. But computer simulation by itself cannot evoke awareness.
C) Appropriate physical action of the brain evokes awareness, but this physical action cannot even be properly simulated computationally.
D) Awareness cannot be explained in physical, computational, or any other scientific terms.

The first one, (A), is what some people refer to as a functionalist point of view. I'm not completely sure whether functionalism might not cover some other things as well, but certainly the term "strong AI" would apply to this view (see Searle 1980). Here, one imagines that mental activity is just a feature of an appropriate type of computation. One is not saying that *any* computation would have mentality, but perhaps those of an appropriate type would. The concepts of consciousness and awareness, and so on, would then be features of computations, in some sense.

Then, there is another point (B), which is, as far as I understand it, more John Searle's point of view (Searle 1980). It states that you could, in principle at least, describe the actions of the brain in terms of computations. That is, if you could have a simulation of the actual brain, that simulation in itself would not have mental qualities. So, he points out that, for example, a simulation of the digestive system would not involve actual digestion. However, in the case of a control system, a "simulation" is good enough, as far as external effects are concerned. We think of the brain as something which controls the body. If the body then behaves like something that is aware, in its *external* aspects, then that is good enough for viewpoint (A). But for (B), we worry about what it *feels*, not just what it *does*.

The distinction here, according to (B), is that even if you behave externally like something which is conscious, it does not mean that you are conscious. Of course, that presupposes that there is a concept of consciousness which is some internal thing which the object itself experiences, in some sense. Much of the philosophical discussion that seems to be going on around this issue seems to deal just with this distinction between (A) and (B).

However, the main point I want to make, and which I have tried to make for some time, is that there is another alternative possibility within scientific explanation. (You might have the view that you should not try to explain mentality in terms of physical things at all, and that somehow the scientific approach is the wrong way to go, but I do not want to talk about (D), which I do not think will be helpful if we want to try to make progress.) The point I want to make is that *scientific explanation or mathematical explanation is not necessarily a computational explanation*. There could well be things in physical action which simply cannot be simulated on a computer. It is certainly a logical possibility that there might be actions which are not simulable on a computer, but whether that is likely to be actually the case in the physical world, as we understand it, is something one has to discuss in some detail. Also, whether it is likely to be relevant to the action of the brain raises a set of questions which I want to come back to a little bit later on.

At this stage, I just want to make clear that there are at least three different viewpoints, and they can be subdivided in different ways, but these are the three scientific viewpoints that seem to be significantly different and talked about. I should just make clear that my own strong preference is with (C). It should be discussed seriously, but it has formerly been ignored almost totally and it is not something that has been properly appreciated.

2.3 What Is Computation?

Now, I should make clear what I mean by computation – what it means in principle. It includes all sorts of things, which people sometimes worry about, and some of them are listed below:

"Computation" (algorithms) include

1) Parallel computers
2) Artificial neural networks (ANN, connectionism)
3) Heuristics
4) Learning
5) Chaotic systems
6) Random ingredients*
7) Interaction with environment*
8) Analogue computers*

In practice, there may be wide differences between some of these things. In principle, a parallel computer is just doing the same thing as a serial computer. It just does things simultaneously instead of one after the other, and that is a matter of detail. And, when I say computation, I do mean the action of a Turing machine, and that really means what a modern general-purpose computer does (see Turing 1950). Technically, a Turing machine is an idealised computer: it has an unlimited storage, it can go on infinitely without wearing out. The essential point is that you can in principle put in more memory if it runs out of it. This is basically what we mean by a computer, what a computer does – it is the action of a Turing machine.

We should understand that computation includes various things which form parts of the operation of a chess playing machine. For example, this could include things that go under the name of "heuristics", i. e., rules of thumb that provide a general guide to assessing positions. Also, learning procedures could be incorporated, such as with Samuel's checkers machine, which does have a learning capability. In all such devices, the procedures according to which that system learns are themselves computational things, as is made clear by the fact that Samuel's learning program for checkers was performed by a computer. We have the general principle: *If it can be simulated on a general-purpose computer, then it is computational.* This should make clear what I mean by the concept "computational". It certainly includes things like chaotic systems; the fact that a process depends critically on the initial state does not mean that it is non-computational.

There is an asterisk on the last three items in the list, which means that technically you might say computation does not include them, but in practice it does. I should make the point clearer here. It could happen that you have a device which involves random ingredients. Now, the way random ingredients are incorporated in calculations, in practice, is usually by using some kind of chaotic calculation. The thing is that you use something which is in effect random, even though technically speaking it is the result of some computation where the output depends critically on the precise input, and you have no way of really distinguishing it from a random behaviour. So, if you have device which actually involves random ingredients, it would not in practice behave differently from something which is a computational device, but which involves some kind of chaotic calculations as part of its behaviour. These points could be elaborated on more, but it is not really appropriate here.

2.4 Environment

One point that should be stressed here is the input from the environment. You could imagine trying to construct an intelligent robot, letting it walk around in the environment, and so on. Does that make it non-computational? The critical issue is whether the environment itself is something you could in principle simulate computationally. You could imagine computationally simulating the entire behaviour of this robot walking around in its environment, provided that the environment is something which can be simulated computationally. Now, if the environment involves chaotic elements, this is okay too. Consider the weather, for example. You do not expect that your robot, or whatever it is you are trying to simulate (you might imagine simulating an entire community of robots), would really depend critically on what the *actual* weather is. All that one needs is that simulated weather be a *plausible* weather. A typical environment, rather than an actual one, would be something which you could simulate.

On the other hand, it might be that the environment has in it some non-computational ingredients. It might be, in particular, that in the training of the robot it is essential that there be input from a human being in this training, and it might be that human beings are, in some sense, non-computational. If that is the case, if you believe that is the way things are, then in a sense, you are already conceding the issue (C). In that case you would not be going along with (A) or (B). Note that my arguments are equally to (A) as to (B). I am not making a distinction. I think most of the discussion people make in philosophical arguments is whether there is a distinction between (A) and (B) – whether it is important, for consciousness, what things are made of, if you like. I think it is important what a thinking machine is made of, but for a different reason. If it is not constructed appropriately, then this gets it outside the restrictions of a computational system. What I want to stress here is that the environment is only giving you something non-computational if you accept that there is something in the physical world which is beyond being simulated effectively in a computational way.

So, if you believe that the outside world can behave non-computationally, then that already means you are accepting the main possibility included in (C). Analogue computers lead to the same thing, basically. An analogue computer acts using some *physical* process, as opposed to just carrying out its action according to the digital procedure of a Turing machine. The normal distinction between analogue computers and digital computers is that analogue computers involve continuous physical parameters, like with slide-rules. The normal view is that any physical parameters could be simulated by digital approximation. If you believe that there is something in the physical world which cannot be simulated by digital computation, then again this is conceding the main argument against (C).

2.5 Neural Networks

In the following, I shall argue that the physical world does not behave computationally, but I should make it clear that "computational" includes *neural networks* and other things of this nature. The term "neural networks" is somewhat awkward, however, because it is used in two different senses. Sometimes people use it for processes in electronic devices of some sort, sometimes they use it for something that is going on in the brain. It is an assumption that these are the same thing, but one does not know. But, what is believed by many people to be going on in the brain at least provides a motivation for what are called *artificial neural networks*. These try to mimic the structure of the brain, in the way that neurons send signals along axons and stimulate the actions of other neurons. This incorporates some kind of mechanism which changes the strength of various connections between neurons at a specific scheme, such as proposed by Donald Hebb many years ago. This is a particular example of a computational mechanism where you have A and B connected and the connection is strengthened if the firing of B tends to follow the firing of A (Fig. 2.2). So, there is some mechanism of this sort, which you could model by a computational device. There is nothing non-computational in this kind of procedure.

Fig. 2.2. Two cells are connected via a synapse, which can be strengthened by a learning procedure

Of course, there might be something else going on in changing the strengths of these connections, and that really is an issue I want to address: what is it that changes the strength in these connections? I shall come back to that later on. If you do have some computational mechanism like the one above, you could simulate that completely computationally and then you have a device like those that are called artificial neural networks. Of course, they are normally technical devices, which are not even trying to be realistic in their ways of modelling the brain. They have rules governing how these connections are strengthened or weakened that is geared to some particular process, such as for building a device that recognises faces, or whatever it might be. This would be a device which learns by experience. However, it is still a device which is entirely computational in the way you set up these so-called neurons and connections, or synapses. They are represented entirely in terms of computational processes that can be put on a general purpose computer. So they are things which are included in what I mean by computational.

2.6 Are Mental Processes Non-computational?

There are, in particular, three things that I want to discuss here. 1) First of all, I should say something about the reasons, most important reasons it seems to me, for believing that there is something in our mental processes, at least in our conscious mental activity, which requires something of a non-computational nature. So, I want to give at least an outline of the arguments involved here. A lot of arguments depend upon the famous theorem by Gödel (1931). I just want to give a very quick indication of how this is used.

2) Suppose that one accepts that there is something non-computational going on, something which cannot be simulated computationally, according to (C). Then you have to see if it is true that the physical world behaves, at least sometimes, in a non-computational way. Well, we do not know for sure, but if we expect to find something non-computational in physical action, it seems that *we have to go outside present day physics*. The case seems very strong to me, that present day physics is something which could be computationally simulated. So, it is a bit unconventional if you take the line that there is something non-computational in physical actions. Some people would say that then you are really going outside physics, but I would say that you are only going outside presently known physics, but *staying within the procedures of scientific explorations*. So, I am certainly not taking the line that science is incompetent to address this issue. I would take a view, however, that present day science is incompetent in this, so one has to go outside present day science. I want to make that more plausible later on.

3) Then, the final point is, do we believe that going outside present day science has a relevance to brain action? I want to indicate where I think the most plausible place is where we can expect to find this kind of thing. I want to get back to this later on.

2.7 The Gödel–Turing Argument

Now, to the argument by Gödel. As I am giving it, it's Turing's argument basically – although Turing was stimulated by Gödel (Turing 1937; 1950). One basic point is that proper computation is not just arithmetic, it involves logical operations, as well. This is in part what one means by computation.

Here is an example, which certainly involves logical operations even though it may look as though it does not. Suppose you have to find the smallest number that is not the sum of three square numbers. "Number" here means a natural number, which is 0, 1, 2, 3, 4, 5, etc. The computation might proceed like this: you start with 0 and try with square numbers from 0 and upwards. There is a very simple minded way of doing this, but I give it just to show that it is a computation. You have 3, 4, 5, and so forth, but when you come to 7 you come to a halt. Seven is the answer, for 7 is indeed the smallest number which is not sum of three square numbers.

Suppose now, you want to find the smallest number that is not the sum of four square numbers. You then find that you can go on forever. This is actually a theorem by Lagrange, proved in the 18th century. It says that every number can be expressed as the sum of four square numbers. Knowing this, we realise that this particular computation will simply go on forever.

An easier non-stopping computation is one that I hope is obvious to everyone: find an even number that is not the sum of two odd numbers. I hope that everyone realises that this too is a particular computation that will never come to an end.

The question is whether mathematicians use algorithmic procedures when they convince themselves that certain computations do not terminate. I should stress here that I am not talking about inspiration, or about originality. I am not talking about how Lagrange may have discovered his argument to show that every number is the sum of four square numbers. It is about how, in principle, you could follow that argument and convince yourself that its conclusion is true. We are really not talking about inspiration, or how mathematicians may come upon their results, but about how we understand.

2.8 Understanding

Understanding is a key concept here. Is understanding itself something which is computational or is it not? The case I shall make is that whatever understanding is – and I am not trying to say what it is, for I do not know what it is – *whatever* understanding is, it is something which is outside computation. That is the key point.

The essential argument can be outlined as follows. Consider computations that act on natural numbers. (An example would be: find a natural number that is not the sum of n square numbers. This is a computation that depends upon the natural number n, i. e., it *acts on* the number n.) Such computations can be listed, $C_0, C_1, C_2, C_3, \ldots$. (This is just a numerical ordering of computer programs.) When acting on a particular n, we have a listing of specific computations, $C_0(n), C_1(n), C_2(n), C_3(n), \ldots$. Now, let us imagine that we have some computational procedure A, which correctly ascertains that certain of these computations do not stop. That is, if $A(p, n)$ successfully terminates, then the computation $C_p(n)$ does *not* terminate. Next, consider the particular computation $A(n, n)$. This acts on the single natural number n, so it must be one of the list of $C_p(n)$, say $C_k(n)$:

$$A(n, n) = C_k(n).$$

Moreover, we have:

"if $A(n, n)$ stops, then $C_n(n)$ does not stop".

Putting $n = k$:

"if $A(k, k)$ stops, then $C_k(k)$ does not stop".

But from above, $A(k, k)$ must be the *same* as $C_k(k)$. It follows that $A(k, k)$ cannot stop, that is, $C_k(k)$ cannot stop. Knowing ourselves, that A is a correct computational procedure for ascertaining that (certain) computations do not stop, we, ourselves, conclude that $C_k(k)$ cannot stop, yet the specific algorithm A is not, itself, powerful enough to ascertain this fact. Since this applies to any algorithm A *whatever*, that we can know correctly ascertains that certain computations do not stop, we conclude:

Mathematicians do not use a knowably correct algorithmic procedure to ascertain mathematical truth.

The question thus arises: what procedures do mathematicians actually use? Let me list some possibilities:

1) an unknowable algorithm, X
2) an unsound algorithm, Y
3) an ever-changing algorithm, Z
4) random elements
5) the environment
6) mysticism
7) non-algorithmic physics

I think it is worth just mentioning here what Gödel and Turing said themselves about this particular issue. Turing seems to have been in camp (2), while Gödel seems to have been in camp (6). I give some approximate quotes here. Turing said something like, "If a machine is to be infallible, it cannot also be intelligent". This is basically possibility (2) above. Certain points should be made here concerning this. One is, that we are not talking about inspiration. We are not talking about how mathematicians actually go about doing their work, discovering or proving theorems. But mathematicians make claims that certain things are true; although they may have made mistakes in their guesses, in coming to these conclusions, ultimately they make clear-cut assertions. Certainly, mathematicians make mistakes, but these mistakes are correctable. They could see, eventually, that slips are wrong. These are correctable mistakes. This does not evade the argument I gave. The point is that truths found by the mathematical community are reliable, and when seen to be wrong they are correctable. It is not that there is something *inbuilt* that is wrong in the way that they proceed.

One can also argue against (2) from the point of view of natural selection. What did Gödel think? He made some comments which lead one to believe that he thought that there was something in the mind that went beyond physics. However, what he said about what we could actually *prove* was that it might be possible to have a *theorem-proving machine* which is in fact equivalent to mathematical intuition, but it could not be proved to be so, nor could it be seen to yield only correct theorems of finitary arithmetic.

These are really two separate ways of circumventing the argument, if you like. Turing says, OK, maybe mistakes are essential to the "mathematicians' algorithm". Gödel argues that, although this putative "mathematicians' algorithm" (i. e., his "theorem-proving machine") could be, in principle, free of errors, there is no humanly knowable way of being sure of this fact. I should make the distinction, moreover, that whereas Turing seemed to believe in this "algorithm", Gödel did not, seeing it only as an implausible logical possibility. Indeed, I believe that one can make quite strong plausibility arguments against this possibility of such a "theorem-proving machine".

2.9 The Robot Society Argument

Perhaps I just should make a few comments about this, although I am not able to give the details here. These are to be provided fully in my new book *Shadows of the Mind* (Penrose 1994), where extensive arguments against all the above possibilities (1), ..., (6) are also given. Suppose you imagine, as do some AI people, that you could make a robot, which can do mathematics. This could mean a robot evolved by natural selection processes and it can involve random ingredients, chaotic ingredients, learning systems. It could be trained by human beings. It could be one of many robots together, and they could interact with each other. However, the entire history of this robot society could be simulated by a (knowable) computation.

The body of mathematical theorems (say, in finitary arithmetic), that the robots could establish as unassailably valid, can be singled out from this computation and provide, for the human programmers, a "robot-theorem-proving machine". Its validity could be known to the humans, but (by Gödel's argument, now from the robots' point of view) not to the robots – thus establishing that such a robot community could never achieve a human-level degree of mathematical understanding.

This argument (given in full detail in *Shadows of the Mind*) enables us to go beyond those of Turing and Gödel. It renders viewpoints (A) and (B), as given above, highly implausible as ever being able to provide explanations of human conscious understanding. The question now arises: what physical processes of a *non*-computational nature might underlie conscious brain action?

2.10 Non-computational Physics

I wish to make a comment about mathematical things which are not computational. There are, indeed, *many* structures in mathematics which are not computational. The question is whether any of these things is relevant to the physical world. Some famous non-computational examples are listed below.

Some mathematical problems without an algorithmic solution:

1) Solubility of Diophantine equations (integer solutions required), e. g., $z^3 - y - 1 = 0$, $yz^2 - 2x - 2 = 0$ (in which case the answer is "yes", with $x = 13$, $y = 7$, $z = 2$, for example).
2) Topological (or smooth) equivalence problems for 4-manifolds (of possible relevance to quantum gravity).
3) Tiling problem: can one cover the plane with a given finite set of distinct polygons? (For example: "yes" for the set consisting of just a triangle; "no" for the set consisting of a regular pentagon and a regular heptagon).

The first one (1) concerns a set of equations and a certain number of unknowns. Everything appearing in the equations is to be an integer.

The question is: can you solve the equations or not? In the particular example shown, the answer is: Yes, you can. But the general problem is, if you get a set of algebraic equations, do they present any integer solutions or not? Can you find a computational procedure which correctly answers *yes* or *no* to all problems of this nature? And the answer is: No, there is no such computational procedure. (It was proved by Yuri Matiyasevich; see Devlin 1988; Davis 1978.) So, solving problems of this nature is something where you, at least from time to time, have to do things which cannot be encapsulated in a computational algorithm. So, there is something going outside algorithmic action, even in problems as simple as this.

The next example, (2), concerns the topological equivalence problem for 4-manifolds (Geroch & Hartle 1986), and is probably the most physically relevant; but I shall not say more about that here. The third example (3) is called the Tiling problem (see Grunbaum & Shepard 1957): can one cover the plane with a given finite set of distinct polygons? To answer yes or no to this question is something that cannot be computationally answered.

The point I want to make is that there might be something about the physical world that depends on solutions to problems like this. One could imagine a "toy universe" described by sets of Diophantine equations. The time-evolution of this "universe" could depend on whether the set of equations providing the "state of the universe" at any one moment has any solution or not. This "universe" would be deterministic, but not *computably* deterministic.

Now, let us ask if in the *actual* physical world there are things which are not computational. In Fig. 2.3, I provide an overview of present-day physics, giving the relationship between classical physics and quantum physics. At the large-scale *classical* level of phenomena, although strong mathematical theorems are still lacking, it seems clear that behaviour at this level is totally computable. If you know the state at one time, you can work out, computationally, the state at any time later.

Now, there is another level of action, the small-scale *quantum* level. There you have the Schrödinger equation, which tells you what happens. It is actually much clearer that the physics is computable down at this level, because the Schrödinger equation is a linear thing – despite the common view that the quantum world is merely probabilistic. Thus, both the classical *and* the quantum levels are, individually, computable, in the sense that there is a computational procedure which describes

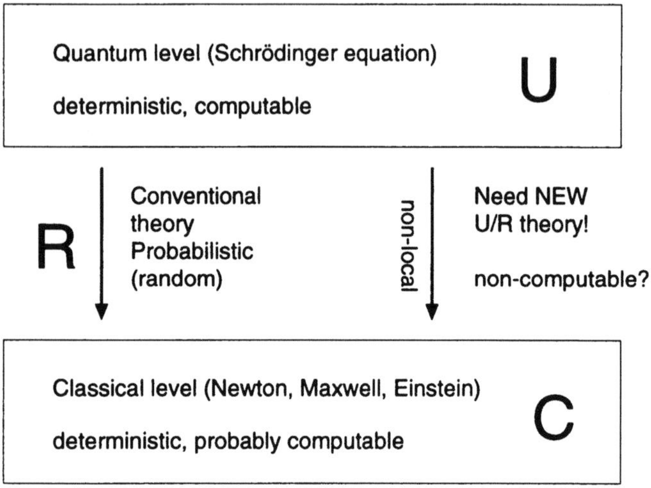

Fig. 2.3. The relation between the classical and quantum levels

what happens. But these two levels individually do not describe all of physics. If you just take the state-vector, or the wavefunction, which describes the quantum level activity, that does not always tell you what happens in the world when you *magnify* something from a small quantum-level scale to larger classical-level scale. Such a magnification is what is called a "measurement", or, if you like, an "observation". But this involves bringing in some new procedure which is not incorporated in the Schrödinger equation, (which is referred to as "U" in Fig. 2.3), or in the classical equations (referred to as "C"). We need, in addition, the procedure which is referred to as *state-vector reduction* ("R" in Fig. 2.3), or *collapse of the wavefunction*. In conventional theory, this is entirely a probabilistic procedure (first arrow in Fig. 2.3).

The conventional theory is very unsatisfactory, since it oscillates between these two levels, all the time bringing in these probabilities when you go from the quantum level to the classical one. I think that is not a good thing for a self-respecting theory. You need something more than this. I think that the new theory we need will be non-computational. Physics is computational at the quantum level (it is just the Schrödinger equation). It seems to be computational up to the classical level, but it is entirely random when you go from one level to the next. Such action can be described as a Turing machine with a randomiser.

However, it seems to me that this is the place where you have to find *new physics*. (This new physics is indicated by the second arrow in Fig. 2.3). Physics is unsatisfactory as it stands, and we just need some new theory. (This requirement is independent of anything I said before.) Some physicists disagree with this, but from my point of view it is clear that something is indeed needed. In most physical systems you can get away with this conventional random procedure, but it may well be that in very delicate organised systems, you will see that it is *not* random. In any case, this involves non-localities. We know that, even with the standard procedure, state-vector reduction is non-local. However, my own conjecture is that this is where *non-computational behaviour* comes in.

2.11 Non-computational Brain Activity

Finally, I want to discuss some new ideas about the level at which non-computational behaviour takes place in brain action. This differs from what was described in the *Emperor's New Mind* (Penrose 1989). The question is: What is actually going on when going from the quantum level to the classical level? Conventional theory is very unclear about this. It just says that this occurs when we "make an observation", but what does that mean? What I am saying is something quite different: that there is a physically defined process which takes place when a quantum system produces effects at the classical level. An example can be given here, in which there is a photon going through a half-silvered mirror, or else it is reflected from the mirror. Quantum theory says that the photon state is a *superposition* of going through and of being reflected. If it goes through, it moves a mass from one place to another. If it is reflected, it leaves the mass alone. The question is, how big does that mass have to be before you consider the state as being "either this or that", as opposed to being in a quantum-level *superposition* of states. The claim (or suggestion) is that you consider the energy E that it would take to displace one instance of this mass in the gravitational field of the other instance of the mass. The reciprocal in Planck units, of that energy gives you a *time* ($h/E = t$), and that time t is the time it takes for the state to jump from the superposed state to one or the other. You find that nucleons (neutrons and protons) will get time scales of hundreds of millions of years, but for tiny droplets of water you find time scales which are more relevant to biological processes.

What is the relevance of this to brain action? In the *Emperor's New Mind* (Penrose 1989) I remarked that nerve transmission does seem to be a classical thing, and whether a nerve fires or not should not be considered as a quantum superposition. But the question is, what about the cytoskeleton activities that Hameroff et al. (1988) has reported about? The cytoskeleton spans each nerve cell and somehow controls the connections between these cells. These microtubules, which are part of the cytoskeleton, are down at the size where quantum-level effects are important. There are cytoskeletal fibres that go right into axons and dendrites, and the cytoskeleton seems to control, or at least be responsible for maintaining the strengths of the synaptic connections. However, one has to look deeper to see what it is that really controls the activity of a neuron and its synapses. Maybe it is the case that the neuron system behaves like a computer, but, if so, it is a computer which all the time is changing, and these changes are governed by deeper-level activity. At the deeper level, you have to ask what the cytoskeleton does, and the case seems to me very strong that there is some kind of quantum coherence that can be involved in signals along these microtubules. However, one needs to have some kind of global coherence across large regions of the brain, so that you can have some kind of global coherence involved in order for quantum mechanics to be important and, more importantly, for the *non-computability*, that I conjecture occurs at the quantum-classical borderline, to have the necessary importance in conscious brain action.

References

Davis, M. (1978). "What is a computation?" in *Mathematics Today; Twelve Informal Essays*, ed. by L. A. Steen, Springer-Verlag, New York

Delvin, K. (1988). *Mathematics: The New Golden Age*, Penguin Books, London

Geroch, R. & Hartle, J. B. (1986). "Computability and physical theories", *Found. Phys*, 16: 533

Gödel, K. (1931). "Über formal unentscheidbare Sätze der Principia Mathematica und verwandter Systeme I", *Monatshefte für Mathematik und Physik*, 38: 173–98

Grünbaum, B. & Shepard, G. C. (1987). *Tilings and Patterns*. Freeman, New York

Hameroff, S. R., Rasmussen, S., Mansson, B. (1988). "Molecular automata in microtubules: basic computational logic of the living state?" in *Artificial Life*, SFI Studies in the Sciences of Complexity, ed. by C. Langton, Addison-Wesley, New York

Penrose, R. (1989). *The Emperor's New Mind*, Oxford University Press, Oxford

Penrose, R. (1994). *Shadows of the Mind*, Oxford University Press, Oxford

Searle, J. (1980). "Minds, brains and programs", *The Behavioral and Brain Sciences*, Vol. 3, Cambridge University Press, Cambridge, MA

Turing, A. M. (1937). "On computable numbers, with an application to the Entscheidungsproblem", *Proc. Lond. Math. Soc.* (ser. 2) 42: 230–265; a correction 43: 544–546

Turing, A. M. (1950). "Computing machinery and intelligence", *Mind* 59, No. 236

3. The Berry Paradox

Gregory J. Chaitin

In early 1974, I was visiting the Watson Research Center and I got the idea of calling Gödel on the phone. I picked up the phone and called and Gödel answered the phone. I said, "Professor Gödel, I'm fascinated by your incompleteness theorem. I have a new proof based on the Berry paradox that I'd like to tell you about." Gödel said, "It doesn't matter which paradox you use." He had used a paradox called the liar paradox. I said, "Yes, but this suggests to me an information-theoretic view of incompleteness that I would very much like to tell you about and get your reaction." So Gödel said, "Send me one of your papers. I'll take a look at it. Call me again in a few weeks and I'll see if I give you an appointment."

I had had this idea in 1970, and it was 1974. So far I had only published brief abstracts. Fortunately I had just gotten the galley proofs of my first substantial paper on this subject. I put these in an envelope and mailed them to Gödel.

I called Gödel again and he gave me an appointment! As you can imagine I was delighted. I figured out how to go to Princeton by train. The day arrived and it had snowed and there were a few inches of snow everywhere. This was certainly not going to stop me from visiting Gödel! I was about to leave for the train and the phone rang and it was Gödel's secretary. She said that Gödel was very careful about his health and because of the snow he wasn't coming to the Institute that day and therefore my appointment was canceled.

And that's how I had two phone conversations with Gödel but never met him. I never tried again.

I'd like to tell you what I would have told Gödel. What I wanted to tell Gödel is the difference between what you get when you study the limits of mathematics the way Gödel did using the paradox of the liar, and what I get using the Berry paradox instead.

What is the paradox of the liar? Well, the paradox of the liar is

"This statement is false!"

Why is this a paradox? What does "false" mean? Well, "false" means "does not correspond to reality." This statement says that it is false. If that doesn't correspond to reality, it must mean that the statement is true, right? On the other hand, if the statement is true it means that what it says corresponds to reality. But it says that it is false. Therefore the statement must be false. So whether you assume that it's true or false you then conclude the opposite! So this is the paradox of the liar.

Now let's look at the Berry paradox. First of all, why "Berry"? Well it has nothing to do with fruit! This paradox was published at the beginning of this century by Bertrand Russell. Now there's a famous paradox which is called Russell's paradox and this is not it! This is another paradox that he published. I guess people felt that if you just said the Russell paradox and there were two of them it would be confusing. And Bertrand Russell when he published this paradox had a footnote saying that it was suggested to him by Mr. G. G. Berry. So it ended up being called the Berry paradox even though it was published by Russell.

Here is a version of the Berry paradox:

> "the first positive integer that cannot
> be specified in less than a billion words".

This is a phrase in English that specifies a particular positive integer. Which positive integer? Well, there are an infinity of positive integers, but there are only a finite number of words in English. Therefore, if you have a billion words, there's only going to be a finite number of possibilities. But there's an infinite number of positive integers. Therefore most positive integers require more than a billion words, and let's just take the first one. But wait a second. By definition this integer is supposed to take a billion words to specify, but I just specified it using much less than a billion words! That's the Berry paradox.

What does one do with these paradoxes? Let's take a look again at the liar paradox:

> "This statement is false!"

The first thing that Gödel does is to change it from "This statement is false" to "This statement is unprovable":

> "This statement is unprovable!"

What do we mean by "unprovable"?

In order to be able to show that mathematical reasoning has limits you've got to say very precisely what the axioms and methods of reasoning are that you have in mind. In other words, you have to specify how mathematics is done with mathematical precision so that it becomes a clear-cut question. Hilbert put it this way: The rules should be so clear, that if somebody gives you what they claim is a proof, there is a mechanical procedure that will check whether the proof is correct or not, whether it obeys the rules or not. This proof-checking algorithm is the heart of this notion of a completely formal axiomatic system.

So "This statement is unprovable" doesn't mean unprovable in a vague way. It means unprovable when you have in mind a specific formal axiomatic system *FAS* with its mechanical proof-checking algorithm. So there is a subscript:

"This statement is unprovable$_{FAS}$!"

And the particular formal axiomatic system that Gödel was interested in dealt with 1, 2, 3, 4, 5, and addition and multiplication, that was what it was about. Now what happens with "This statement is unprovable"? Remember the liar paradox:

"This statement is false!"

But here

"This statement is unprovable$_{FAS}$!"

the paradox disappears and we get a theorem. We get incompleteness, in fact. Why?

Consider "This statement is unprovable". There are two possibilities: either it's provable or it's unprovable.

If "This statement is unprovable" turns out to be unprovable within the formal axiomatic system, that means that the formal axiomatic system is incomplete. Because if "This statement is unprovable" is unprovable, then it's a true statement. Then there's something true that's unprovable which means that the system is incomplete. So that would be bad.

What about the other possibility? What if

"This statement is unprovable$_{FAS}$!"

is provable? That's even worse. Because if this statement is provable

"This statement is unprovable$_{FAS}$!"

and it says of itself that it's unprovable, then we're proving something that's false.

So Gödel's incompleteness result is that if you assume that only true theorems are provable, then this

"This statement is unprovable$_{FAS}$!"

is an example of a statement that is true but unprovable.

But wait a second, how can a statement deny that it is provable? In what branch of mathematics does one encounter such statements? Gödel cleverly converts this

"This statement is unprovable$_{FAS}$!"

into an arithmetical statement, a statement that only involves 1, 2, 3, 4, 5, and addition and multiplication. How does he do this?

The idea is called Gödel numbering. We all know that a string of characters can also be thought of as a number. Characters are either 8 or 16 bits in binary. Therefore, a string of N characters is either $8N$ or $16N$ bits, and it is also the base-two notation for a large positive integer. Thus every mathematical statement in this formal axiomatic system

"This statement is unprovable$_{FAS \leftarrow}$!"

is also a number. And a proof, which is a sequence of steps, is also a long character string, and therefore is also a number. Then you can define this very funny numerical relationship between two numbers X and Y which is that X is the Gödel number of a proof of the statement whose Gödel number is Y. Thus

"This statement is unprovable$_{FAS}$!"

ends up looking like a very complicated numerical statement.

There is another serious difficulty. How can this statement refer to itself? Well you can't directly put the Gödel number of this statement inside this statement, it's too big to fit! But you can do it indirectly. This is how Gödel does it: The statement doesn't refer to itself directly. It says that if you perform a certain procedure to calculate a number, this is the Gödel number of a statement which cannot be proved. And it turns out that the number you calculate is precisely the Gödel number of the entire statement

"This statement is unprovable$_{FAS}$!"

That is how Gödel proves his incompleteness theorem.

What happens if you start with this

"the first positive integer that cannot
be specified in less than a billion words"

instead? Everything has a rather different flavor. Let's see why.

The first problem we've got here is what does it mean to specify a number using words in English? – this is very vague. So instead let's use a computer. Pick a standard general-purpose computer, in other words, pick a universal Turing machine (*UTM*). Now the way you specify a number is with a computer program. When you run this computer program on your *UTM* it prints out this number and halts. So a program is said to specify a number, a positive integer, if you start the program running on your standard *UTM*, and after a finite amount of time it prints out one and only one great big positive integer and it says "I'm finished" and halts.

Now it's not English text measured in words, it's computer programs measured in bits. This is what we get. It's

"the first positive integer that cannot
be specified$_{UTM}$ by a computer program
with less than a billion bits".

By the way the computer program must be self-contained. If it has any data, the data is included in the program as a constant.

Next we have to do what Gödel did when he changed "This statement is false" into "This statement is unprovable." So now it's

"the first positive integer that can be proved$_{FAS}$
to have the property that it cannot
be specified$_{UTM}$ by a computer program
with less than a billion bits".

And to make things clearer let's replace "a billion bits" by "N bits". So we get:

"the first positive integer that can be proved$_{FAS}$
to have the property that it cannot
be specified$_{UTM}$ by a computer program
with less than N bits".

The interesting fact is that there is a computer program

$$\log_2 N + c_{FAS}$$

bits long for calculating this number that supposedly cannot be calculated by any program that is less than N bits long. And

$$\log_2 N + c_{FAS}$$

is much much smaller than N for all sufficiently large N! Thus for such N our *FAS* cannot enable us to exhibit any numbers that require programs more than N bits long. This is my information-theoretic incompleteness result that I wanted to discuss with Gödel.

Why is there a program that is

$$\log_2 N + c_{FAS}$$

bits long for calculating

"the first positive integer that can be proved$_{FAS}$
to have the property that it cannot
be specified$_{UTM}$ by a computer program
with less than N bits" ?

Well here is how you do it.

You start running through all possible proofs in the formal axiomatic system in size order. You apply the proof-checking algorithm to each proof. And after filtering out all the invalid proofs, you search for the first proof that a particular positive integer requires at least an N-bit program.

The algorithm that I've just described is very slow but it is very simple. It's basically just the proof-checking algorithm, which is c_{FAS} bits long, and the number N, which is $\log_2 N$ bits long. So the total number of bits is, as was claimed, just

$$\log_2 N + c_{FAS}.$$

That concludes the proof of my incompleteness result that I wanted to discuss with Gödel.

Over the years I've continued to do research on my information-theoretic approach to incompleteness. Here are the three most dramatic results that I've obtained:

1) Call a program "elegant" if no smaller program produces the same output. You can't prove that a program is elegant. More precisely, N bits of axioms are needed to prove that an N-bit program is elegant.

2) Consider the binary representation of the halting probability Ω. Ω is the probability that a program chosen at random halts. You can't prove what the bits of Ω are. More precisely, N bits of axioms are needed to determine N bits of Ω.
3) I have constructed a perverse algebraic equation

   $$P(K, X, Y, Z, \ldots) = 0.$$

   Vary the parameter K and ask whether this equation has finitely or infinitely many whole-number solutions. In each case this turns out to be equivalent to determining individual bits of Ω. Therefore N bits of axioms are needed to be able to settle N cases.

These striking examples show that sometimes you have to put more into a set of axioms in order to get more out. (2) and (3) are extreme cases. They are accidental mathematical assertions that are true for no reason. In other words, the questions considered in (2) and (3) are irreducible; essentially the only way to prove them is to add them as new axioms. Thus in this extreme case you get out of a set of axioms only what you put in.

How do I prove these incompleteness results (1), (2) and (3)? As before, the basic idea is the paradox of "the first positive integer that cannot be specified in less than a billion words." For (1) the connection with the Berry paradox is obvious. For (2) and (3) it was obvious to me only in the case where one is talking about determining the **first** N bits of Ω. In the case where the N bits of Ω are scattered about, my original proof of (2) and (3) (the one given in my Cambridge University Press monograph) is decidedly not along the lines of the Berry paradox. But a few years later I was happy to discover a new and more straight-forward proof of (2) and (3) that is along the lines of the Berry paradox!

In addition to working on incompleteness, I have also devoted a great deal of thought to the central idea that can be extracted from my version of the Berry paradox, which is to define the program-size complexity of something to be the size in bits of the smallest program that calculates it. I have developed a general theory dealing with program-size complexity that I call algorithmic information theory (*AIT*).

AIT is an elegant theory of complexity, perhaps the most developed of all such theories, but as von Neumann said, pure mathematics is easy compared to the real world! *AIT* provides the correct complexity concept for metamathematics, but not necessarily for other more practical fields.

Program-size complexity in *AIT* is analogous to entropy in statistical mechanics. Just as thermodynamics gives limits on heat engines, *AIT* gives limits on formal axiomatic systems.

I have recently reformulated *AIT*.

Up to now, the best version of *AIT* studied the size of programs in a computer programming language that was not actually usable. Now I obtain the correct program-size complexity measure from a powerful and easy to use programming language. This language is a version of *LISP*, and I have written an interpreter for it in *C*. I am writing a book employing this new approach that is entitled *The Limits of Mathematics*.

So this is what I would have liked to discuss with Gödel, if I could have spoken with him now.

Finally, I would like to suggest a list of questions for future research.

- Find questions in algebra, topology and geometry that are equivalent to determining bits of Ω.
- What is an interesting or natural mathematical question?
- How often is such a question independent of the usual axioms? (I suspect the answer is "Quite often!")
- Show that a classical open question in number theory such as the Riemann hypothesis is independent of the usual axioms. (I suspect that this is often the case, but that it cannot be proven.)
- Should we take incompleteness seriously or is it a red herring? (I believe that we should take incompleteness very seriously indeed.)
- Is mathematics quasi-empirical? In other words, should mathematics be done more like physics is done? (I believe the answer to both questions is "Yes.")

References

Books by the Author

Chaitin, G. J. (1990). *Information, Randomness & Incompleteness*, second edition, World Scientific, Singapore

Chaitin, G. J. (1992). *Algorithmic Information Theory*, fourth printing, Cambridge University Press, Cambridge

Chaitin, G. J. (1992). *Information-Theoretic Incompleteness*, World Scientific, Singapore

Recent Papers by the Author

Chaitin, G. J. (1993). "On the number of n-bit strings with maximum complexity", *Applied Mathematics and Computation* 59: 97–100

Chaitin, G. J. (1993). "Randomness in arithmetic and the decline and fall of reductionism in pure mathematics", *Bulletin of the European Association for Theoretical Computer Science*, No. 50 (June 1993), pp. 314–328

Chaitin, G. J. (1993). "Exhibiting randomness in arithmetic using Mathematica and C", *IBM Research Report RC-18946*, 94 pp., June 1993

Chaitin, G. J. (1993). "The limits of mathematics—Course outline & software", *IBM Research Report RC-19324*, 127 pp., December 1993

Chaitin, G. J. (1994). "Randomness and complexity in pure mathematics", *International Journal of Bifurcation and Chaos* 4: 3–15

Chaitin, G. J. (1994). "Responses to 'Theoretical Mathematics ...' ", *Bulletin of the American Mathematical Society* 30: 181–182

Chaitin, G. J. (1994). "The limits of mathematics (in C)", *IBM Research Report RC-19553*, 68 pp., May 1994

Books and Articles by Other Authors

Davis, M. (1978). "What is a computation?", in L. A. Steen, *Mathematics Today*, Springer-Verlag, New York

Rucker, R. (1982). *Infinity and the Mind*, Birkhäuser, Basel

Tymoczko, T. (1986). *New Directions in the Philosophy of Mathematics*, Birkhäuser, Basel

Rucker, R. (1987). *Mind Tools*, Houghton Mifflin, Boston

Pagels, H. R. (1988). *The Dreams of Reason*, Simon & Schuster, New York

Berlinski, D. (1988). *Black Mischief*, Harcourt Brace Jovanovich, New York

Herken, R. (1988). *The Universal Turing Machine*, Oxford University Press, Oxford

Stewart, I. (1989). *Game, Set & Math*, Blackwell, Oxford

Boolos, G. S. and Jeffrey, R. C. (1989). *Computability and Logic*, third edition, Cambridge University Press, Cambridge

Ford, J. (1989). "What is chaos?" in Davies, P. *The New Physics*, Cambridge University Press, Cambridge

Casti, J. L. (1989). *Paradigms Lost*, Morrow, New York

Nicolis, G. and Prigogine, I. (1989). *Exploring Complexity*, Freeman, New York

Casti, J. L. (1990). *Searching for Certainty*, Morrow, New York

Küppers, B.-O. (1990). *Information and the Origin of Life*, MIT Press, Cambridge, MA

Paulos, J. A. (1991). *Beyond Numeracy*, Knopf, New York

Brisson, L. and Meyerstein, F. W. (1991). *Inventer L'Univers*, Les Belles Lettres (English edition in press)

Barrow, J. D. (1991). *Theories of Everything*, Oxford University Press, Oxford

Ruelle, D. (1991). *Chance and Chaos*, Princeton University Press, Princeton, NJ

Nørretranders, T. (1991). *Mærk Verden*, Gyldendal, Copenhagen

Gardner, M. (1992). *Fractal Music, Hypercards and More*, Freeman, New York
Davies, P. (1992). *The Mind of God*, Simon & Schuster, New York
Barrow, J. D. (1992). *Pi in the Sky*, Oxford University Press, Oxford
Hall, N. (1992). *The New Scientist Guide to Chaos*, Penguin, Harmondsworth
Reichel, H.-C. and Prat de la Riba, E. (1992). *Naturwissenschaft und Weltbild*, Hölder-Pichler-Tempsky, Vienna
Stewart, I. (1992). *The Problems of Mathematics*, Oxford University Press, Oxford
Dewdney, A. K. (1993). *The New Turing Omnibus*, Freeman, New York
Çambel, A. B. (1993). *Applied Chaos Theory*, Academic Press, New York
Svozil, K. (1993). *Randomness & Undecidability in Physics*, World Scientific, Singapore
Casti, J. L. (1994). *Complexification*, Harper Collins, New York
Gell-Mann, M. (1994). *The Quark and the Jaguar*, Freeman, New York
Nørretranders, T. (1994). *Verden Vokser*, Aschehoug, Copenhagen
Wolfram, S. (1994). *Cellular Automata and Complexity*, Addison-Wesley, New York
Calude, C. (–) *Information and Randomness*, Springer-Verlag, in press

4. Mind as Phenotype

Robert Rosen

4.1 Introduction

"Phenotype" is a uniquely biological concept. Etymologically, it derives from the Greek verb *phainein*, to show; the same root from whence derives the word *phenomenon*.

"Phenotype" is a rather recent word, but "phenomenon" is old. The dictionary tells us that "phenomenon" means "any fact, circumstance or experience that is apparent to the *senses*, and that can be *scientifically* described or appraised". That is, phenomena devolve on *cognition*, and hence on direct (or sometimes indirect) *sensory* experience. From this follows another characterization of "phenomenon", typically regarded as synonymous with the one we have given: "the *appearance* or *observed* features of something experienced, as distinguished from reality or the thing *in itself*". Indeed, the profound limitations of cognition as an exclusive guide to this reality, this *Ding an sich*, have been emphasized by philosophers from Plato to Kant.

In biology, "phenotype" thus pertains to what we can see or measure about an organism, or about its behaviors, or its properties; its anatomy, and its physiology. It thus comprises whatever there is about organisms which directly exposes them to cognition in the first place, and that which renders such knowledge sufficiently *objective* for scientific purposes.

But in biology, for the first time in science it becomes apparent that "phenotype" is not enough; that the objective cognition of such visible properties does not and cannot suffice to make biology a *science*. That is why the term "phenotype" comes paired with a dual concept, *genotype*, which pertains not directly to phenomena themselves, but rather to their causal bases. In other branches of science, such as standard physics and chemistry, there was deemed no need for any such distinction; genotype and phenotype simply *coalesce* in them, united in a common world of phenomena. But with biology, they do not coalesce any more; they have somehow *bifurcated*. Indeed, one of the fundamental problems of approaching even phenotypes *reductionistically*, exclusively in terms where genotype and phenotype do coincide, resides precisely here. We shall look at such questions, from a variety of viewpoints, as we proceed.

If biology raises such questions already, they are raised even more vividly by considerations of mind and sentience or consciousness. The fact is that each of us is more aware of these than we can possibly be of any mere cognition or sensory impression. The awareness is thus *phenotype*, but it has nothing to do with cognitions of phenomena, which indeed presuppose it. Above all, it is *subjective*, and this subjectivity is intensely contagious, spreading to everything it touches; volition, intentionality, imagination, emotion, and all the rest of it, including cognition, and mathematics, and science.

In the present work, we shall explore two quite different ways of attempting to "objectify" the subjective phenomena of mind and sentience; i. e., of pulling them into the cognitively based world of phenomena and phenotype. First, there is a strategy we shall call *mimesis*. It says, basically, that an understanding of any property resides entirely in finding some purely mechanical device which mimics the property "sufficiently closely". We can then *impute*, from the mimicry itself, the causal basis of the behavior from the mimic to what it mimics; a tacit invocation of Occam's Razor. In particular, if we can find such a machine which mimics in its behaviors some aspect of "intelligence" or "thought", then we can say the *machine* already "thinks". Thus, understanding that mechanical mimic. In effect, we are turning the mimicry relation around, and extrapolating from the machine to anything which *mimics the machine*. This, of course, is the substance of the familiar *Turing Test*, about which we shall have more to say later. We can already see, however, that arguments of this kind teeter on the brink of *equivocation*.

A second approach is loosely called *Reductionism*. The basic idea here is to devolve any phenotypic property of a material system onto those manifested by a sufficiently extensive class of material subsystems; e. g., chemical molecules, atomic particles, subatomic particles, and points south. These subsystems are themselves required to be *context independent*, in the sense that (a) they must be fractionable, or separable from any larger system to which they belong, and (b) their own properties do not depend on where they came from or how they were isolated. A phenotypic property of such a larger system is "objective" to the extent that it is expressible entirely in terms of those of such constituents.

The approach here is thus poles apart from mimesis; instead of relating material systems through their capacity to "behave alike", independent of what they are made of, we rather relate systems, however diverse in phenotype, through their reducibility into such context-independent elements. Stated another way; any phenotypic property of any material system is but another way of perceiving, or of apperceiving, the very same properties of the elements that we can cognize directly about them. In particular, a property like sentience, a subjective property of certain material systems, is but a reexpression and reformulation of those same cognitive properties of electrons, protons, neutrons, and what have you, which we perceive directly about them when they are in isolation. To use a philosophic term, any phenotypic character of any material system is already *immanent* in a few properties of a few kinds of context-independent reductionistic elements. It is precisely this "immanence" which makes the procedure universal.

Each of these pictures of objectivity, considered by itself, seems persuasive enough. The trouble is that they do not coincide; indeed, they are not even compatible with each other. One picture (the mimetic one) pertains to phenotype alone; the other (the reductionistic one) pertains to genotypes. If they are to be co-extensive, the two must indeed coincide. But in organisms, as we have seen, they do not coincide. Moreover, any world in which they do coincide (e. g., the world of *algorithms*) *is a barren, sterile one; far too impoverished to do mathematics in, let alone support anything resembling life, let alone mind and sentience.*

A specific consequence of the reductionistic picture we have sketched is this: that one must not expect to learn anything new about reductionistic units (e. g., about electrons, or protons, or neutrons) by looking at phenotypes of larger systems, like organisms. To suggest otherwise is to strike a blow at reductionism itself, and its own presumed identification with objectivity; those who believe in reductionism see it as an advocacy of *Vitalism*. Be that as it may; the thesis of our argument below is precisely that: if life, and sentience are phenotypic, then *physics* is not what we presently think it is. And likewise, even if they are not phenotypic.

4.2 The Turing Test

The familiar Turing Test actually asserts several things. On the one hand, it proposes an "effective" procedure for discriminating phenotypes into two classes, which we may call *intelligent* and *unintelligent*. It thus proposes this "intelligence" as itself an objective phenotypic property of something; as objective in its way as the charge on an electron. On another hand, it asserts the non-existence of any such procedure for discriminating between this "intelligence" and what we subjectively call *thought* (and, by extension, the entire subjective realm of *psyche*). On these grounds, it proposes a direct identification of the former with the latter. And finally, it proposes that any systems manifesting this phenotypic property of "intelligence", being to that extent objectively indiscriminable, can simply be interchanged; any one of them is a surrogate for any other.

Thus, according to your taste, the Turing Test is either a great step forward, or a great step backward. If you believe the former, it constitutes a kind of objectification of thought or *psyche* itself, bringing it within the realm of hard scientific investigation. If the latter, it constitutes a tissue of equivocations, just as the closely related Church's Thesis turned out to be. Complicating the issue is that the Turing Test adduces "evidence", produced by closely mimicking the very same kinds of cognitive procedures which produce "evidence" in general, about anything. Thus, of something is wrong with the Turing Test, it casts doubt upon unthinking reliance upon such "evidence" in general.

Turing framed his initial argument within a very small world; a world of *machines*. This is the world of algorithms, consisting of finite-state hardware executing algorithms embodied in programs or software. Together, the hardware and the programs process inputs, or data, to generate corresponding outputs. In this little world, the *behaviors* so generated, embodied in the relations between inputs and their corresponding outputs, are precisely what constitutes *phenotype*.

As we have noted, some of these phenotypes can be effectively discriminated as being "intelligent". The actual discriminator may itself consist of such a programmed device (originally, it was supposed to be a human being, but it subverts the whole point of the argument to keep something subjective in this role). The discriminator proceeds by supplying input data ("questions") and evaluating the corresponding outputs, or "answers". In the simplest terms, the classifier is comparing these "answers" to what its own would be to those same "questions". It will thus classify another machine as "intelligent" or not to the extent that it *simulates the discriminator*. Or more accurately, simulates it "enough", based on a short list of such "questions".

Hence, in this world, the objective phenotypic property of "intelligence" is identified with "sufficiently simulating" the system which determines the property. Clearly, any such discriminator will always, vacuously, classify itself as "intelligent"; i. e., as itself possessing the phenotype in question.

Now this same discrimination procedure, though initially posed in a small world of hardware and software, is not restricted to that world; the test can be applied to other material systems (e. g., to organisms, or to human beings). Indeed, *any* system to which the test can be applied at all can be said to possess the phenotypic property the test objectifies; i. e., to be intelligent or unintelligent, depending on the outcome.

In any case, the test provides us with a presumably well defined class of systems, possessing a certain objective phenotypic character ("intelligence"), and this class will have at least some "machines" in it (e. g., the one which actually makes the discrimination). This kind of classification of systems in terms of phenotypic characters or behaviors constitutes the basis for any *taxonomy*, and rests entirely on such evidence supplied by discriminators or meters for the characters in question. Indeed, it is generally regarded as *information* about something to specify a taxonomic class to which it belongs (e. g., it is information about an organism to say that it is a mammal; it is information about a particle to say that it is an electron). In the same way, it is information about a machine program to say that it is "intelligent".

Indeed, precisely this kind of taxonomic information provides one way of characterizing what a physicist would call the *state* of a system under investigation. He would, from this viewpoint, have to say that such a state always comprises, or expresses, the sum total of all such taxonomic features, and is simply a specification of the intersection of every taxonomic class to which it belongs. Unfortunately, that is not the only way of talking about state; there is also a quite different, reductionistic way, and as we have already noted, the two are quite different. We shall get to these matters subsequently.

The next point to recognize is that we also know, as objectively as anything, the profound limitations and inadequacies of algorithms and machines; i. e., of the universe in which Turing originally framed his test. Especially is this so when one such machine is attempting to assess the behaviors (phenotypes) of another (as in, e. g., the halting problem, posed by Turing himself). These considerations lead us to expect that taxonomic classes based on phenotypic properties like "intelligence", must contain many other things besides machines (which, in themselves, constitute but another taxonomic class). And if so, it brings into serious question whether it is legitimate at all to extrapolate from one *member* of such a taxonomic class to another, simply on the basis of "evidence" that they share one such class.

Indeed, to the extent that the Turing Test is itself a mimic of traditional experimental, cognitive procedures for classifying phenotypes, it casts serious doubt on those very procedures as a universal, "objective" way of characterizing material systems in general. It certainly provides no basis for asserting any kind of relation of *surrogacy*, or of *modeling*, between them. In other words, the Turing Test ends up by casting doubt upon reliance on any circumscribed set of cognitive modes for this purpose. Accordingly, far from its initial intent to "objectify" thought by mechanizing it, the Test only reinforces and magnifies the need for something other than phenotypic information as a basis for science itself. In the last analysis, what is missing from such pictures is *genotype*.

4.3 Reductionism

If there is one thing which has been made clear by the history of biology, brief as it has been, it is that phenotypes entail little or nothing about their associated genotypes. Thus there is nothing in the study of anatomy and physiology, nor for that matter in development or evolution (i. e., in morphogenesis) which even obliquely suggests anything like, e. g., DNA. In fact, quite the contrary; the assertion that diverse kinds of systems can still be phenotypically indiscriminable, as the Turing Test claims, clearly argues against any such thing. That is partly why, in biology, genotype and phenotype are so different.

On the other hand, procedures like the Turing Test are in themselves based on a very limited view of what an experiment or observation of a system consists of. They cannot comprehend, for example, experiments involving taking a hammer to a machine's hardware, nor putting its software through a shredder. Such experiments radically change everything about a system which the Turing Test assumes preserved; they change the state set, the dynamics, the "transfer function" of the system, in extreme and irreversible ways. Input-output experiments must not be allowed to do any of these things; they must not change the dynamics on which the input/output relation, the "transfer function", depends; otherwise, the experiments would be worthless. For this reason, part of the very characterization of a "machine" asserts the independence of its behaviors (phenotypes) from the experiments used to probe it; namely, that the execution of a program must not change either the hardware or the program being executed. In molecular biology, this taboo is the content of what Crick long ago called the "Central Dogma", that phenotypes must not affect genotype, or that "information" must not flow from the former to the latter. This Dogma is essential in maintaining the mechanical perception of cells which molecular biologists like. But in fact, this kind of Dogma has a far wider currency, which has gone far to determining the very character of physics itself.

Destructive experiments, which inherently violate all these *caveats*, serve to create entirely new *systems* from given ones, with new state sets and new dynamics, generally very different from the system we began with. That is to say, such destructive experiments drastically change the original system's phenotypes, its genotypes, and the relations between them. That is, they destroy everything on which mimesis is based. The *only* relation between the original system and the new one, is the fact that they are both comprised of the same material; the same "stuff" (and even that is not always true). To borrow a term from chemistry, what survives destructive experiments are, at best, relations of *isomerism*.

Thus, an initial intact organism is *isomeric*, in this sense, isomeric to a cell population obtained by passing it through a sieve; and each of these cells is isomeric to a spectrum of chemical fractions residing in a hundred different test tubes; and each of these is isomeric to a cloud of electrons, protons, neutrons, or what you will. Each of these isomers is thus treated as a surrogate organism, not because it *behaves* like an organism, but precisely (and only) because it is an isomer of one.

Unlike mimetic strategies, which seek surrogates for organisms via common phenotypic behaviors, reductionism seeks surrogates in a class of ever-bigger isomers, and seeking to express the phenotypes of our original system in terms of those of a big-enough isomer. The ultimate goal is to find a canonical set of "units" such that *every* material system (e. g., an organism) is isomeric to some population of them, and such that every *property*, every phenotype of such a system can be "effectively" expressed in terms of those of the units.

It is not my intention here to provide a detailed critique of this reductionistic strategy; it would take a long time, and I have done it elsewhere. Suffice it to say that its conceptual base is exceedingly thin, and that it has never been of direct conceptual use in dealing with the basic problems of biology (let alone of mind and sentience). On the other hand, it has been exceedingly fruitful in the deployment of procedures for isomerising things, and in the generation of data about them. The actual pertinence of these activities, however, remains essentially a matter of faith; it is in itself a subjective judgment.

Suffice it to say here that the spirit of reductionistic approaches is very much like that of canonical form theorems in mathematics, many of which are variants of the familiar Jordan Canonical Form for matrices. In all these theorems, you need to know only a limited number of elementary units, and a limited number of ways to put them together; from these, one generates enough "canonical forms" so that everything in the universe of discourse differs from one of the canonical ones only by a subjective thing like a transformation of co-ordinates. For instance, a great deal of quantum mechanics rests on the Spectral Theorem, which builds (Hermetian) operators or observables canonically out of projections, and which is in fact a paraphrase of the Jordan Canonical Form to situations which are not finite-dimensional. The interpretation of these projections in terms of observational procedures (in quantum mechanics inherently destructive) is well known; hence the idea that every property of everything is comprehended within it.

On the other hand, canonical form theorems are useless unless one can specify the transformations which turn the properties of a canonical isomer, in which we are generally not interested, into those of another one in which we are. The very existence of these is primarily a matter of faith, let alone that they are in any sense "computable", as Church's Thesis requires. It should be noted that the destructive experiments on which reductionisms depend are inherently irreversible, in any material sense, yet the effects of these presumed transformations is precisely to reverse them; to go backwards from a big population of elementary units to a particular material system (e. g., an organism) of which they are an isomer.

Many years ago, I heard a routine of Woody Allen which bears on exactly this point. As he told it, he acquired a Rolls-Royce while in England, and wanted to return with it to the US. On the other hand, he didn't want to pay the duty on it. So he hit on the idea of disassembling it, packing the parts into many suitcases, and describing them to the customs inspectors as "modern sculpture", not dutiable as art. He was successful, got his many suitcases home, and proceeded to try to reassemble his car. In his first attempt, the parts yielded 200 bicycles. On the second attempt, he got many lawnmowers. And so it went; he never could retrieve the car.

4.4 Some Necessary Reassessments

As we have seen, there have been two distinct and highly inequivalent strategies utilized for approaching things like the "mind-brain" problem, and hence to express subjective things in presumably "objective", scientific terms. The first is to replace the system of interest by some behavioral or phenotypic mimic; something which manifests some of its behaviors, but not in general the causal underpinning of those behaviors. Then something like Occam's Razor is invoked to argue that explaining these behaviors in the mimic is adequate for explaining them in the original system. The second strategy is to seek some canonical material isomer, which now behaves completely differently (i. e., has a different phenotype), but such that every original behavior is a transform of something about that isomer. Indeed, the point of seeking the canonical isomer is to reveal more clearly its context-independent units or constituents; in these units, genotype and phenotype appear to coalesce. The properties of the units themselves are few in number, and are themselves known cognitively, and are entirely independent of any larger isomer to which they might belong.

In each case, however, we replace the actual system of interest by some kind of surrogate, and end up studying that surrogate. It is in this surrogate that "objectivity" resides, in one form or another, and which is imputed to the original system. But of course, as we have seen, the basis for this surrogacy is completely different in the two strategies.

We have discussed the inequivalence of these strategies at great length in other places. We have shown, for example, that the first strategy attaches "state sets" to a system, which are direct sums of its own phenotypic properties, while the second regards state sets as direct products of the units in some canonical isomer. Indeed, in effect, the coalescence of genotype and phenotype, to which we have alluded several times, amounts to mandating the equivalence of direct sums and direct products; something which is almost never true. Correspondingly, any world in which it is true must be a very special world, a *simple* world (cf. Rosen 1991).

Since these two approaches are of such different generality, neither entails the other, nor are they generally inter-transformable. Hence, each one by itself is seriously incomplete as a universal scientific strategy.

As we have stressed, the inequivalence of these strategies becomes of central importance in biology, where it is literally incarnated in the inequivalence of phenotype and genotype. We might say that this inequivalence is in itself a phenotypic property manifested by the class of material systems we call organisms; it constitutes as good a diagnostic of "the living state" as there is. From the beginning, I have started from this separation between genotype and phenotype; it is the very point of departure of the relational models I have called (**M**, **R**)-systems (Rosen, *loc. cit.*). It also indicates why organisms resist reductionism so mightily, since in a sense the whole point of reduction is to dispense with that very property.

For such reasons, I have argued for a long time that life poses the most serious kinds of challenges to physics itself; orders of magnitude more serious than the ones electrodynamic phenomena posed for Newtonian mechanics, or that spectra and chemical bonding posed for 19th-century classical physics. More specifically, the expectation that phenomena of life or mind could be assimilated directly into physics as merely a minor technical bubble, of no conceptual significance, were mistaken. In a certain sense, the problem is that contemporary physics has never, of its very nature, had room for a concept of genotype. It can only maintain its present form by staying close to material systems in which genotype can be ignored; in biology we meet material systems where it cannot be ignored any more. But this requires a radical conceptual restructuring of physics itself, from the ground up.

If one wishes to call the incorporation of genotypic concepts into physics a kind of *complementarity*, *à la* Bohr, there is no objection. But this does not help physics itself actually do it.

At heart, the concept of genotype addresses the causal basis of phenomena, or phenotypes; phenotypes are *effects*. What I am suggesting is that these causal bases are themselves not just other phenomena. Hence, in particular, that phenotypic things like life and mind are not just another way of perceiving, or apperceiving, the small repertoire of cognitively visible, context-independent properties of a few reductionistic units. Nor can the phenotypes be separated from their causal bases, as is tacit in mimetic approaches like the Turing Test.

These problems occupy center stage in biology. But of course they have antecedents and predecessors in contemporary physics itself. They are there treated as technical anomalies, raising no issues of principle, just as was, say, the anomalous specific heat of the diamond, or the behavior of the perihelion of Mercury. Only in hindsight could it be recognized that such anomalies were foretokens of revolutionary upheavals, rather than technical difficulties which could be resolved with just more cleverness in applying received principles. So, for example, most physicists dismiss "the measurement problem" in quantum mechanics as something of no importance, even though it is, in a sense, an embryonic form of the mind-brain problem itself.

The problem here is that the very acquisition of data, the very cognition of phenomena (phenotype) in a material system, requires you to consider a larger system ("system + observer") and not to smaller ones, as reductionism (or context-independence, or "objectivity") requires. This in turn creates a chicken-egg situation, an *impredicativity*; specifically, you must know the larger system to characterize the smaller, but you cannot know the larger until the smaller is characterized. Contemporary physics does not like impredicativities in Nature, but of course it is deeply concerned with preserving the "objectivity" of the cognitive processes (measurements) on which it bases itself. Pais (1991), in his assessment of this problem, declares it only a game played by mathematicians and epistemologists, devoid of impact on *practical* physics. But then, practical people tried to refute the Zeno paradoxes by getting up and walking around the room.

Nevertheless, the excluded process of finding out about a given system by putting it into a *larger* system (i. e., by creating a definite *context* for it), in the teeth of the impredicativities this generates, is perhaps the only exit from the boxes which reductionism alone, or mimesis alone, create for us. Biology already forces out of these boxes, for life and mind are not to be found within them.

The pervasive fear of impredicativities, and hence the attempts to encompass all of "objective" reality within purely predicative bounds, is quite needless. To be sure, one must be careful with them; they can lead to paradoxes if one is not. But biology clearly shows that one cannot simply dispense with them, in physics any more than in mathematics.

Indeed, if one tries to capture something like the genotype-phenotype dualism in purely predicative (i. e. algorithmic or computable) terms, or what turns out to be the same thing, to identify direct sums with direct products, one generally requires an infinite number of these algorithms. No canonical forms, of the type described in the preceding section, exist for systems of this type, which I have elsewhere termed *complex*. In effect, contemporary physics is based on the belief that *no complex systems exist in Nature*. This is but another form of Church's Thesis, and I have argued at great length that it is false. On the other hand, if it is false, then there are plenty of material systems (organisms) whose phenotypes can neither be regarded as a mere re-cognition of a limited set of properties of reductionistic units, nor exhausted by a limited number of mechanical mimics. Either of these commands a fundamental enlargement of our physical perceptions of material nature itself.

4.5 Some Remarks on "Inertia" and "Gravitation"

The fact that effects do not in general entail their own causes, i. e., that phenotypes do not entail their own genotypes, is central to the entire discussion above. If this is so, then merely cognizing the phenomena which populate our "objective" world of events and behaviors, is not enough; it is necessary, but almost never sufficient. One way of expressing the difficulties of both reductionism and mimesis is the underlying belief that "enough" necessary conditions comprise a sufficient one.

Thus, neither of the following is true: that "enough" cognitions of brain will suffice to make a mind, or that "enough" behaviors of mind will suffice to make a brain. The belief that there is always a threshold, embodied in the word "enough", which will make such assertions true, is in its turn a consequence of reliance on repetition of a limited repertoire of rote processes, and the belief that "enough" such repetitions will take you to any threshold. Repetition of rote processes is of course the essence of algorithm. And as we have seen, algorithms embody a world without impredicativity, a world in which genotype and phenotype coincide.

In such a world, analysis and synthesis are simple inverse processes. There is no difference in describing a system in terms of how its own behavior is determined by its subsystems, or what that system does to larger systems to which it might belong, and if there is no difference, we might as well concentrate on the former. This is part of the predicative view, that we must only look to subsystems to understand something, and never try to understand it by looking to larger systems, or specific *contexts*. As we have seen, this not only creates impredicativities, but seems to deny "objectivity" or context independence itself.

Therefore it is instructive to hark back to the foundations of classical Newtonian mechanics, a kind of quintessence of objective science, and in particular at how it treats (a) the *response* of a particle to an (external) force impressed upon it, and (b) how that same particle can *exert* a force upon other particles.

The problem (a) is addressed, in one fashion or another, by Newton's Three Laws of Motion. The problem (b), on the other hand, requires a completely separate and independent Law of Universal Gravitation. The latter has nothing to do with the former; it cannot be entailed from Laws of Motion, nor they from it.

Now how a mechanical particle responds to a force is tied to its *inertia*, and is embodied in a specific parameter called *inertial mass*. On the other hand, how that same particle *generates* a force is embodied in a conceptually quite different parameter, called *gravitational mass*. The former is phenotypic in nature; the latter is genotypic.

Now it turned out that, though these parameters are very differently defined, on the basis of entirely separate laws, the *values* of these parameters were numerically the same. This strange fact was later seized upon by Einstein; he regarded it, not as a happy coincidence, but rather an expression of something very deep about physics itself. He ultimately expressed it in terms of a *Principle of Equivalence* between inertia and gravitation, asserting the non-existence of any experiment, done entirely within a mechanical system, which could discriminate between the one and the other; this ended up as a cornerstone of General Relativity. In our terms, it asserts the fundamental identity of genotype (gravitation) and phenotype (inertia) in mechanics.

But even here, this stops being true once you try to interpret, or *realize*, Newtonian impressed forces as arising from the gravitation of other particles; i. e., as soon as you put your particle into a bigger system of particles. Roughly, when you do this, you are allowing that impressed force to be determined (at least in part) by the particle itself. The particle is now (phenotypically) behaving in a "field of force" which it participates in generating. Although a classical particle is forbidden from (gravitationally) pushing itself when it is alone, adding other particles allows it to push things which can push it. The resultant impredicativities, which in fact plague all field theories of particle-based forces, arise as soon as you try to identify inertia (phenotype) with gravitation (genotype) independent of any larger context.

This is why, for instance, one cannot solve a three-body problem reductionistically, by solving two-body and one-body problems.

The impredicativities we have described have bothered physicists for centuries, because even here, they involve an intertwining of genotypic and phenotypic considerations, whose absolute separation (or absolute identification) is at the heart of the entire picture. They deeply disturbed Ernst Mach, for example, and led him to propose a view of mechanics quite different from, and incompatible with, the Newtonian, based entirely on phenotypes. In practice, most physicists ignore such problems, through the employment of "mean-field approximations" which serve to again isolate the inertia of a particle from the effects of its own gravitation.

The capacity for "self-forcing", the effects of a particle's gravitational properties upon its own behaviors when there are other particles around, can therefore not easily be expressed by differential rate equations, or dynamical systems, at all. They are, rather, expressed by networks of interlocking differential forms which describe direct and indirect effects on inertia of gravitation; what I have elsewhere (e. g., Rosen 1983) called *activation-inhibition networks*, *agonist-antagonist networks*, etc. These networks of differential forms collapse into a finite system of ordinary rate equations, or equations of motion, when all these differential forms are exact, but not otherwise. It turns out that the non-exactness of some or all of these differential forms is another way of describing *complexity*, as we have characterized it above.

Thus, even in classical mechanical physics, any presumed "objective" identification of inertia and gravitation, or of genotype with phenotype, already raises deep questions bearing on the adequacy of both mimetic and reductionistic strategies for coping with material reality. These problems, and many others like them, become critical in using those strategies to cope with life and with mind.

What has all this to do with the mind-brain problem, our initial point of departure? Basically, I suggest that contemporary physics, the science of material reality, and the touchstone of objectivity rooted entirely in cognition, rests upon a host of presumptions which in themselves are not only noncognitive in nature, but which need not hold (and, I would argue, do not hold) in systems which live and think. There is nothing "unphysical" in acknowledging these facts, nor does it mean an abandonment of the "laws" on which it is presently based; it only means that, in addressing life and mind as material phenomena, which they are, we cannot expect to find only specializations of these "laws", but rather generalizations of them.

I will try to illuminate these last assertions by looking as I have often done, at cognate situations in mathematics. In particular, at what I have called "real" number theory, and its various Formalist syntactic simulacra (which can be called "artificial" number theories; sometimes "constructive" number theories). It is a primary upshot of Gödel's Theorem that the "real" theory is always much bigger than any of the "artificial", predicative, algorithmic, machine-based ones. Nevertheless, the "real" theory does not violate any of the "laws" on which the "artificial" ones are based. But it does not specialize these "laws"; it rather enormously *transcends* them. On the other hand, there is no purely phenotypic *evidence* which can effectively discriminate the "real" from the "artificial". Certainly, this discrimination cannot be made from entirely within such an "artificial" arithmetic alone. To do so requires a radical enlargement of the cognitive base of the system itself, and a corresponding enlargement of their genotypic foundations.

4.6 A Brief Summation

We have proceeded from the fact that nobody doubts, or even *could* doubt, his own life and his own sentience. Nor does anyone doubt that these things are inseparable from, and indeed properties of, a certain material system, which he calls his "body". That makes them, in biological parlance, *phenotype*, even though the knowledge of them is non-cognitive. Nevertheless. they should be amenable to the general procedures used by biologists to study phenotype, and more generally, to the procedures used by physicists to study material systems. But none of these procedures, individually or collectively, has shed much light on either the basis of life or of mind.

So something is wrong. There are two possibilities; either the procedures themselves are universally sound, and we have merely been remiss in carrying them out properly, or the procedures themselves are limited and/or flawed. In the latter case, no amount of industry or diligence based entirely on these procedures will suffice. In the above discussion, I have argued for the latter possibility.

Part of the problem here is that "evidence" alone, short of actually solving these problems, does not suffice; as noted above, such "evidence" supplies at best necessary conditions, and not sufficient ones. Nevertheless, like everyone else, I have throughout the above discussion adduced such "evidence".

An important part of this "evidence" is historical in nature. The fact is that neither reductionistic approaches to our problems, anchored in (theoretical) physics, nor mimetic approaches, have shed any real light on the problems themselves. As to the former, whenever theoretical physics ventures into the biological realm, it either has nothing to say, or what it says has so far been wrong. Historically, whenever this happens, it has bespoken a foundation crisis of the first magnitude in physics itself. And it has already happened many times. Why should it be any different this time?

As we have seen, a *reduction* of biology to contemporary physics throws the problem onto finding some canonical isomer of an organism, and expressing the latter's phenotypes entirely in terms of independently-known, cognitive, observable properties of the units constituting that canonical isomer. The fact that every property of every other isomer must also be so expressible simply sweeps the problem under the rug of "initial conditions". In my view, this simply begs the entire question; a concession of abject scientific impotence.

Nevertheless, the difficulty of finding the "right" needle of initial conditions in the haystack of all possible ones provides the basis for the common view that organisms constitute a rare and negligible specialization of physical universals we already independently know. To suggest otherwise is to expose oneself to charges of Vitalism and worse.

We have suggested above, to the contrary, that organisms vastly generalize the cognitive bases on which contemporary physics itself rests, in the same way that "real" arithmetic generalizes any purely syntactic "artificial" normalization. In that sense, one does learn deep lessons about *physics* by grappling with phenomena of life and of mind.

We have framed our discussion in terms of the distinction in organisms between phenotype and genotype, whereas physical approaches are based on a mandated coincidence or coalescence of the two. Their distinction in organisms is reflected in the fact that no amount of phenotypic evidence entails anything about the underlying genome.

As essential feature of our discussion is the necessity of learning about a material system, not only by taking it apart into subsystems, but also by putting it into larger systems with which it can interact. This was the substance of our discussion of "inertia" and "gravitation" above. In biology, these two ways of viewing a given material system are already radically different, and highly inequivalent. But as we have noted, the invocation of specific larger contexts creates *impredicativities* which contemporary physics does not like to admit; that is precisely why there is a "measurement problem" in quantum theory, and a "life-organism" problem in biology, and a "mind-brain problem". Such problems ultimately force us to re-examine some deeply ingrained habits of thought, which it is entirely possible are bad habits.

The physicist Schrödinger, in his famous essay of 1943, argued forcefully that biology mandated "new physics". Both physicists and, today, molecular biologists, are (for quite different reasons) vehement in their denial of any such thing. But today, fifty years later, when conventional approaches have failed to move us an inch closer to problems of life and mind, such exhortations may begin to seem a bit more plausible.

4.7 Some Concluding Remarks

In the preceding discussion, we have taken an exclusively *synchronic* view of the mind-brain problem; a partition of the present world, now, today, into an "objective" part and a "subjective" one. The former, as we have said, is the province of *physics* in the broadest sense; a world of phenomena or phenotype, in which subjectivities are allowed no causal role. The mind-brain problem, paradoxically enough, seeks to pull mind and awareness, the basis of subjectivity, back into this external objective world.

One powerful motivation for trying to do this (quite apart from the subjective lust for unification) is found not in synchrony, but rather in *diachrony*; in history, or as the biologist would say, in phylogeny. Put briefly, it is possible to imagine a time "before" consciousness; a time "before" life. At such a time, there was *only* an objective, phenomenal world; hence no separation or partition between objective and subjective. The partition itself, according to this view, had to *emerge* diachronically from the pre-existing, abiotic, amental objectivities, through a process of "evolution". Hence, one important aspect of the mind-brain problem lies in running this diachronic process, this history or evolution from abiotic to biotic, backward. Specifically, to try to pull the subjective into the objective by pulling the present into the sufficiently remote past.

Such a diachronic approach is different in spirit from either mimesis or from reductionism, neither of which is diachronic in this sense.

To consider this approach in detail is not possible in this short space. We will only make a few remarks about it, in terms of the analogy we have drawn between "objective" in the causal universe and "formalizable" in the mathematical one. This latter, of course, does not have a diachronic aspect, in the same sense. But it does have a historical one. The fact is that the very concept of formalization was an exceedingly late one; first expressed by Hilbert, but with roots going back to Euclid and before. Hilbert, of course, thought initially that everything in mathematics was formalizable, but Gödel very quickly showed that this was not so; indeed, that formalizability was rare in mathematics. Diachronically, in fact, mathematics ran quite the other way; it did not emerge from "objective", formalizable beginnings, to successively more "complex", nonformalizable elaborations. There was no such process of "evolution" in the history of mathematical thought. If anything, this "evolutionary" process ran quite the other way. In particular, the formalizable or "simple" systems were not intermediate evolutionary stages, or building blocks, in the elaboration of more "complex", non-formalizable mathematical systems. Nor, as Gödel showed, are they even useful subsystems in studying their synchronic properties; their "physiology", if you will.

My suggestion is to take this mathematical metaphor seriously in the material world as well. Namely, that diachronic evolutionary processes do not necessarily go from "simple" to "complex" in matter, any more than they do in mathematics.

References

Pais, A. (1982). *Subtle is the Lord ...*, Clarendon Press, Oxford

Rosen, R. (1983). "Causal Systems which are not Mechanisms" in *Theoretical Biology and Complexity*, ed. by R. Rosen, Academic Press, New York, pp. 165–203

Rosen, R. (1991). *Life Itself*, Columbia University Press, New York

Part II

Sub-cellular Levels

In the previous section it was argued that the current theoretical framework of mathematics, physics, and biology is insufficient to explain all aspects of the mind-matter problem. In particular, explanations of certain aspects of mind, such as understanding, intuition, free will, and subjectivity, should be based on non-computational processes. The important point is that they would thus be impossible to simulate or describe with computational models. The currently rather popular "connectionist" approach to the problem (which will be described in the next block) was challenged, and completely new principles of physics and biology were called for.

Regardless of whether such new principles are needed, it is quite obvious that any theory of mind has to be based on experimental data from neuroscience. If we want to get a deeper understanding of how neural processes of the brain are related to mental processes, we have to take a closer look at the structure and organization of the brain. The question is, how close? Should we stop at the level of networks, neurons, and synapses, or should we go further down in scale and also include sub-cellular levels? Could cognitive functions be critically dependent on single molecules and sub-cellular structures, such as enzymes, ion channels, and microtubules, or even on atoms and electrons? In that case, could perhaps quantum mechanical effects be important?

Neurons embody intricate networks of protein filaments with a multitude of interacting molecules and ions, and could encompass computational and possibly also non-computational processes. These aspects are neglected in most theoretical and modeling work, but may, in fact, hold the key to the mental activity of the brain. This is pointed out by Hameroff and Conrad & Conrad in the following two articles. They stress that it is important to consider the intricate mechanisms governing the molecular structures and functions involved in neural information processing. Such mechanisms could be under quantum mechanical influence, and could provide the physical basis for the non-algorithmic processes that Penrose claims are vital for human consciousness. Could these processes perhaps also allow for a true, (indeterministic) free will, which is considered to be essential for our conscious experience? Indeed, Hameroff and Conrad & Conrad provide suggestions for the physical sites, microtubules and enzymatic processes, respectively, where the interaction between mind and matter could take place.

Hameroff, in his article, gives many reasons why it is inadequate to regard neurons as simple switching elements of a complex network system, and why intracellular networks of microtubules should be taken into account when explaining emergent phenomena, such as cognition. In particular, quantum coherence, triggering conformational changes in the microtubular proteins, could be essential for various aspects of consciousness, including the unitary sense of self, insight, free will and the nature of experience. This notion of a (quantum) link between the cytoskeleton and cognitive functions is supported by experimental evidence and models of microtubular information processing, as given in the article. In an addendum, Hameroff describes the collaborative Penrose–Hameroff model of consciousness ("orchestrated objective reduction" – Orch OR – in microtubules) which grew from the Abisko gathering.

Conrad & Conrad also strongly oppose the simplified computational cognitive, or connectionist, view of neural structure and function. They claim that the physical-chemical-dynamical nature of organisms is inextricably tied to mentality, and base their argument on three "conjectures": 1) the rule-following behavior of organisms includes choice and intention, 2) only carbon-based non-programmable and evolving systems can have the real-time computational capabilities exhibited by organisms, and 3) such systems depend on certain physical features, in particular the collapse of the wave function. One main point here is that the formal maps required to describe everyday behavior of an organism are so complex that they must be inseparably linked to the material nature of the system, i. e., to the organic molecules of biological cells. According to this hypothesis such behavior cannot be duplicated by structurally programmable (silicon) machines, or by any system not using the electronic-conformational interactions of organic (carbon-based) molecules.

It should be noted that Conrad & Conrad do not claim to have any strong evidence in support of their hypothesis. Rather, they infer from the immense complexity of biological systems that any artificial, programmable system could not possibly be able to perform, in real time, the "everyday" rule-based behavior of a living organism. Although this might not be a theoretical impossibility, in practice it would be.

Both the chapters by Hameroff and Conrad & Conrad provide a link from the microscopic, sub-cellular level to the macroscopic level of cognition and behavior. In both chapters, the specific nature of biomolecular structures that should be necessary for mental activities is also stressed. In pointing at the computational powers of neural (cellular and sub-cellular) systems, they also provide a physically oriented framework for Penrose's and Chaitin's more abstract discussions on computational problems related to mind.

We thus come back to the problem of which aspects of mental activity are computational and which are not. Conrad & Conrad distinguishes two types of mentality: cognition and consciousness, where cognition has to do with learning, reasoning, knowing, and remembering, whereas consciousness is described as being aware of one's environment, one's own body, or one's own mental state. Could it be that such cognitive functions are based on computational or algorithmic processes, and thus could be sufficiently described with present-day physics and mathematics? If so, it should be possible, at least in principle, to simulate these processes on a digital computer (if enough time and processing elements were available). The basis for these computations could be at a network level or, as suggested here, at a cellular or sub-cellular level. Mental activities related to consciousness, on the other hand, could be of a non-computational (non-algorithmic) nature, which might require some new physics and (possibly) mathematics. Both Hameroff and Conrad & Conrad believe the key to these activities may be related to quantum-mechanical effects, acting at the molecular level of the neural systems.

5. Quantum Automata in Cytoskeletal Microtubules: A Nanoscale Substrate for Cognition

Stuart R. Hameroff

5.1 Introduction: Global State Emergence

Cognition may be defined as "the emergence of global states in a network of simple components" (Varela et al. 1992). Such emergence may occur through "connectionism" which offers a strategy for self-organized thought and consciousness without a central processor. Connectionist artificial neural networks ("ANNs") can adapt and learn, but the analogy between ANNs and biological systems falters in several areas. First, biological neurons and their synapses are extremely complex and by themselves resemble computers more than fundamental switches. Second, a number of important ANN paradigms require "back-propagation", tuning of input parameters from output conditions. Such feedback in biological neurons would seem to require retrograde information flow. Third, single-cell organisms such as *Paramecium* routinely perform complex adaptive behaviors which would demand significant computational power to emulate, yet do so without benefit of a single synapse! Fourth, some aspects of consciousness are seemingly inexplicable in terms of computer analogies. These include a "unitary sense of self" (the "binding problem"), non-algorithmic processing or "insight" (Penrose 1989), nondeterministic free will, and the essential nature of human experience.

In addition to connectionist networks, global state emergence occurs in quantum physics. "Macroscopic quantum states" such as superconductivity, superfluids, and lasers emerge from quantum coherence among distributed, simple components. Quantum coherence in biological systems has been predicted by Fröhlich (1968, 1975, 1986), and a model of consciousness based on "Fröhlich" quantum coherence among "neural proteins", has been proposed by Marshall (1989). Thus "simple components" in biological cognitive networks should imply molecular events, specifically conformational states of proteins coupled to quantum events. The cytoskeleton, of which microtubules (MTs) are key elements, is a dy-

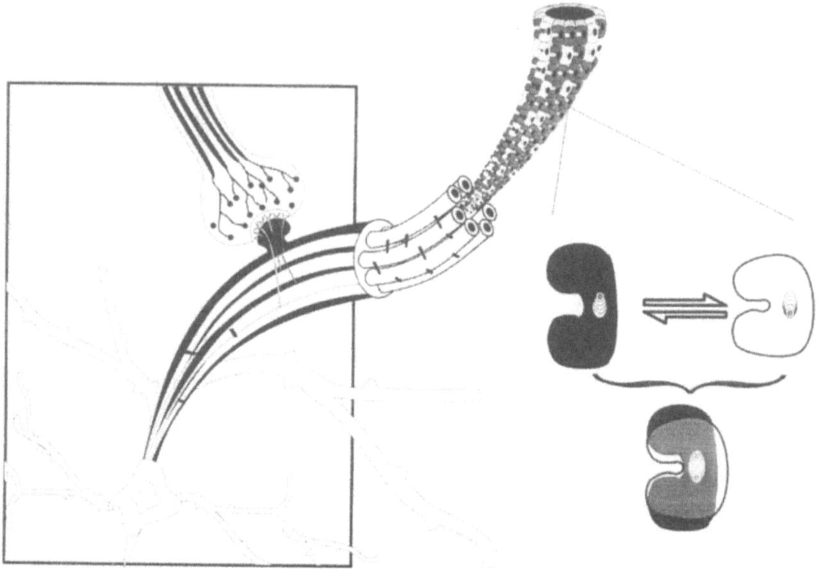

Fig. 5.1. Schematic of quantum effects in cytoskeletal microtubules. *Box at left*: Brain neuron with one enlarged dendrite extending upward, one incoming synapse shown on dendritic spine, cytoskeleton visible within neuron. *Middle*: Bundle of microtubules interconnected by microtubule associated proteins, single magnified microtubule comprised of tubulin subunits in two possible states. *Lower right*: Zoom of single tubulin in two possible states coupled to position of single electron within hydrophobic pocket. *Lower right bottom*: Tubulin in quantum superposition of both states

namical network of protein polymers which organizes interiors of living cells. Cytoskeletal involvement in neural function is becoming clearer (e. g., Hirokawa 1991) and can account for complex functions of neurons, "back-propagation" (Dayhoff et al. 1994) and adaptive behaviors in single-cell organisms (Hameroff 1994). Cytoskeletal information processing increases the brain's theoretical "bit-state" capacity 7–10 orders of magnitude (Hameroff et al. 1992); however, this ultra-reductionism still fails the inexplicable: self, insight, free will, and experience. Only quantum effects approach these questions. In this chapter, possible quantum mechanisms and cognitive capabilities of the cytoskeleton are explored (Fig. 5.1).

5.2 Microtubules and the Cytoskeleton

Eukaryotic cells from single-cell protozoa to cortical neurons are comprised of protoplasm, which in turn consists of membranes, organelles, nuclei, and the bulk interior medium of living cells: cytoplasm. Nineteenth-century light microscopists described cytoplasm as containing or consisting of "reticular threads", "alveolar foam", or "watery soup" (Burnside 1974). Development of the electron microscope through the 1960s did not initially illuminate the substructure of cytoplasm because the commonly-used fixative (osmium tetroxide) was dissolving fine structure and the cell was still often perceived to be a "bag of watery enzymes." In the early 1970s, with the advent of glutaraldehyde fixation, delicate tubular filamentous structures were found in virtually all cell types and they came to be called microtubules (MT). Originally thought to provide merely structural or bone-like support, MT and other filamentous structures such as actin, intermediate filaments, and centrioles were collectively termed the "cytoskeleton" (Fig. 5.2). Recognition that complex, dynamic activities of MT and other cytoskeletal elements were essential for organization, movement, and growth of cellular cytoplasm finally brought the cytoskeleton out of the closet.

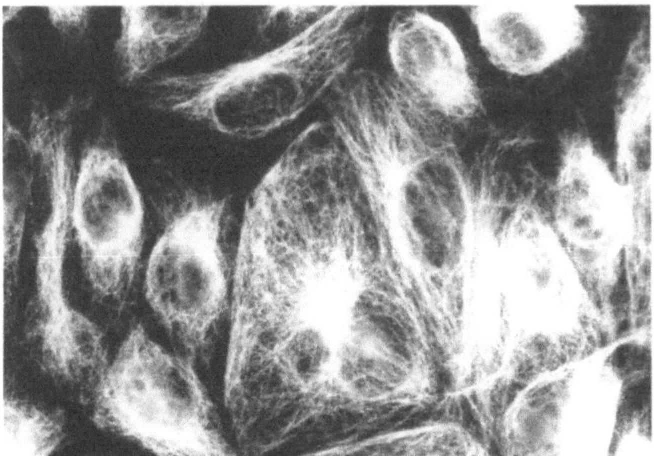

Fig. 5.2. Immunofluorescence micrograph of tissue culture cells stained with anti-tubulin antibody. White, lace-like network comprised of microtubules. Courtesy Lawrence Vernetti

Parallel arrayed MT are interconnected by cross-bridging proteins (i. e., MT associated proteins: "MAPs") to other MT (Fig. 5.3), organelles, and smaller filamentous proteins (actin, intermediate filaments, etc.) to form a dynamic gel whose cellular activities (i. e., mitosis, growth, differentiation, locomotion, exocytosis, phagocytosis, synapse modulation, dendritic spine formation, cytoplasmic movement, neurotransmitter release, etc.) are essential to the living state. Orientation and directional guidance of these cytoskeletal functions depend on centrioles: cylindrical assemblies of nine MT triplets (structurally similar to cilia) arranged in pairs oriented perpendicular to each other (Fig. 5.4). These centriole pairs, which constitute the focal point of the centrosome, or cell center, organize the array of cytoplasmic MT during interphase, duplicate at mitosis to nucleate the two poles of the mitotic spindle, and establish orientation and architecture for the next generation of cells.

Of the various filamentous structures which comprise the cytoskeleton, MT are the most prominent (Dustin 1984) and best characterized, and appear best suited for dynamic information processing. MT are hollow cylinders 25 nanometers (nm) in diameter whose lengths may span meters in certain neurons. MT cylinder walls are assemblies of 13 longitudinal protofilaments which are each a series of subunit proteins known as tubulin. Each tubulin subunit is a polar, 8 nm dimer which consists of two slightly different classes of 4 nm, 55 kilodalton monomers known as α and β tubulin (Fig. 5.4). The tubulin dimer subunits within MT are arranged in a hexagonal lattice which is slightly twisted, resulting in differing neighbor relationships among each subunit and its six nearest neighbors. MT, as well as their individual dimers, have dipoles with negative charges localized toward α monomers (DeBrabander 1982). Thus MT are "electrets": oriented assemblies of dipoles which are predicted to have piezoelectric and ferroelectric properties (Athenstaedt 1974; Mascarenhas 1974; Tuszynski et al. 1995).

Contractile or enzymatic MAPs may be attached to MT at specific tubulin dimer sites; MAP attachments can thus result in various helical patterns on MT surface lattices (Burns 1978; Kim et al. 1986). Conformational changes occur in tubulin (Engelborghs 1992); e. g. one monomer can shift 29 degrees from the dimer's vertical axis (Melki et al. 1989).

Genes for α and β tubulin are complex, multi-gene families which give rise to varying tubulin isozymes. During evolution some tubulins have been highly conserved for basic cell functions; however, extensive MT heterogeneity exists in complex cells due to genetic diversity, expression,

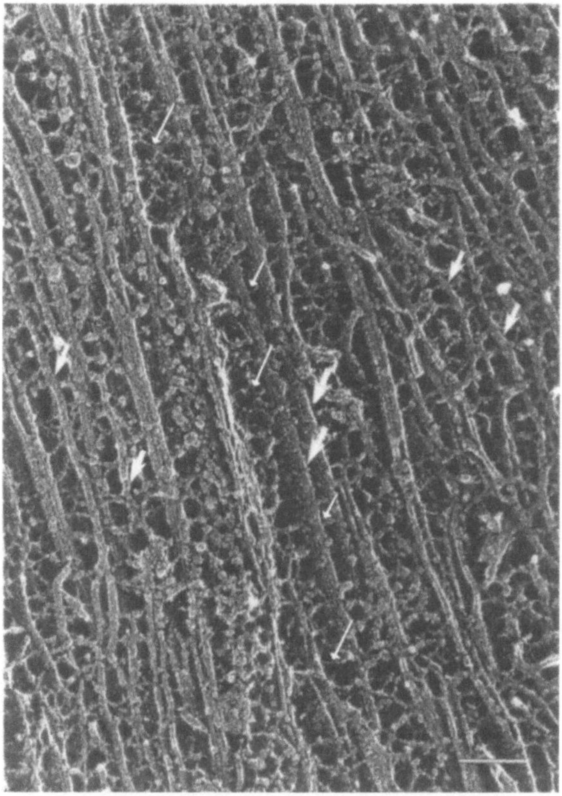

Fig. 5.3. Electron micrograph of quick frozen, deep etched neuroplasm from interior of rat neuron. Frequent short cross-bridges (*short arrows*) are found between microtubules (*thick black arrows*). Longer cross bridges also connect microtubules (*long arrows*). Short black arrows indicated neurofilaments. Scale bar (*lower right*) 100 nanometers. Reprinted with permission from Hirokawa (1991)

post-translational modifications, MAPs, and assembly patterns. For example, two-dimensional gel electrophoresis has shown 17 different varieties of β tubulin exist in mammalian brain MT, whereas fewer exist in other tissues (Lee et al. 1986). Tubulin structure may also be altered by "post-translational" modification: enzymatic alteration (usually addition or removal of amino acids such as glycosylation or detyrosination) triggered by intracellular events including second messenger activities. Post-translational modification of the cytoskeleton may serve to "hard-wire" information into long-term memory.

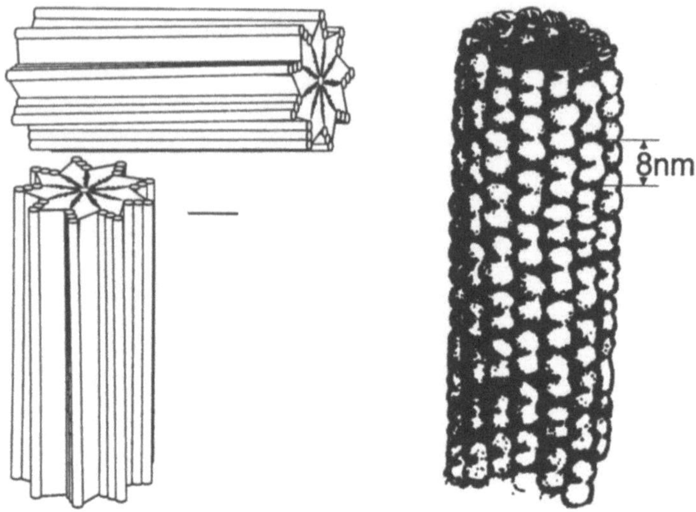

Fig. 5.4. *Left*: The centriole is a perpendicular pair of microtubule (MT) megacylinders. Each megacylinder is comprised of nine microtubule triplets. Scale bar: 100 nanometers. *Right*: Microtubule structure from x-ray diffraction (Amos & Amos 1991). MTs are cylinders 25 nanometers in diameter whose walls are 13 protofilaments, each a string of 8 nanometer tubulin dimers. Alpha and beta tubulin monomers comprise the dimers

MT self-assembly and disassembly are dynamic, complex processes whose states depend on various factors including temperature and calcium ion concentration. Oriented by centrioles, MT polymerisation determines the architecture and form of cells which can quickly change by MT depolymerisation and reassembly in another direction. GTP, an energy-providing analogue of ATP, binds to free, unpolymerised tubulin; GTP-tubulins then self-assemble to form MT in an entropy driven process. Within assembled MT, GTP hydrolyses to GDP, imparting energy into the MT lattice via tubulin conformational changes. MT whose ends are comprised of GTP-tubulin are stable and will continue to grow. MT whose ends are comprised of GDP-tubulin are unstable and will depolymerise rapidly. As GTP is hydrolysed to GDP within MT, GDP-tubulin is exposed at MT ends and, unless stabilized by MAPs, centrioles, or other structures, MT rapidly disassemble. Free GDP-tubulin is reconverted to GTP-tubulin which becomes available for reassembly. *In vitro*, this "dynamic instability" (Kirschner & Michison 1986; Hotani et al. 1992) results in erratic alterations between growing and shrinking phases; *in vivo* "selectionist" activity occurs in which MT networks can adaptively probe or retreat in cellular appendages including axonal and dendritic growth cones and developing synapses (e. g., Reinsch et al. 1991; Sabry et al. 1991). Recent work has shown that MT spatial structures or patterns may form under the maintenance of a sustained energy source (GTP hydrolysis), suggesting that MT may be dissipative structures (Tabony & Job 1992).

Contractile MAPs such as dynein and kinesin participate in ciliary bending as well as intra-neuronal ("axoplasmic") transport which moves material within axons and dendrites and plays a major role in maintaining and regulating synapses. Other MAPs form structural bridges and stabilize MT, preventing their disassembly, and some may be phosphorylated by second messengers. Thus MAP-MT interactions help determine cell architecture and dynamic function (Lasek 1981).

5.3 Cytoskeletal Information Processing

The cytoskeleton has a justified existence in terms of providing structural support and transport. Why invoke additional, unproved responsibilities as the cell's on-board quantum computer and substrate of consciousness? One reason is a lack of satisfactory alternative mechanisms for real-time, intra-cellular organization. The standard explanations of membrane depolarisations, ion fluxes, diffusion of signal molecules, and genetic regulation require faith in an organized soup. The cytoskeleton provides a solid state network which interconnects all neural elements; it is well situated to be the cell's nervous system.

Vassilev et al. (1985) showed that MT can transmit signals between membranes (Fig. 5.5). More recent work has demonstrated propagation of conformational states and vibrational "shimmering" in microtubules (Vernon & Woolley 1995; Yagi et al. 1994; Hunt & Stebbings 1994). Cytoskeletal proteins directly link to nerve membrane ion channels and receptors, and evidence links the cytoskeleton to nerve membrane excitability and synaptic transmission (Matsumoto & Sakai 1979; Hirokawa 1991). Receptor binding of neurotransmitter molecules in neuronal synapses triggers second messenger cascades, which include G proteins, [Ca^{++}], adenyl cyclase, phosphatidylinositol, cyclic AMP, etc. These transducing elements, in turn, respond by activation of dynein ATPases and cytoplasmic phosphatases and kinases such as Ca^{++}-calmodulin protein kinase, protein kinases A and C, etc. (Eriksson et al. 1992).

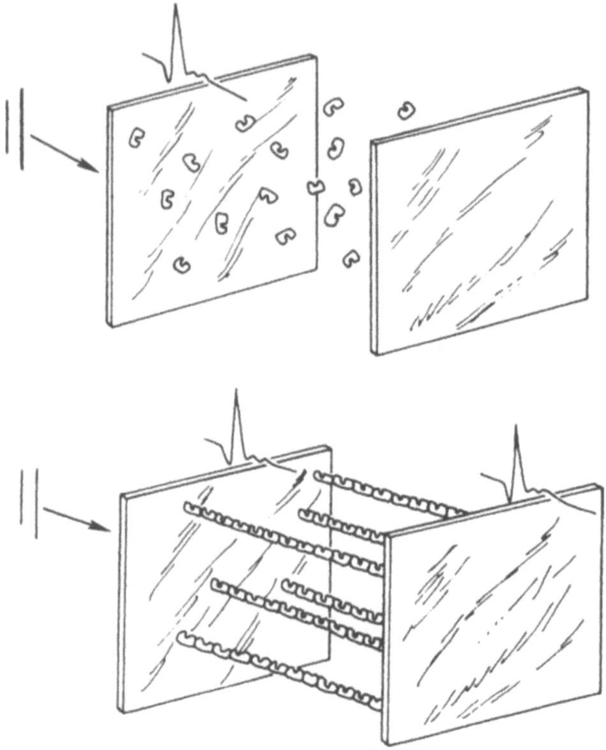

Fig. 5.5. Schematic representation of experiment by Vassilev et al. (1985). Two excitable membranes are suspended in parallel millimeters apart in a buffer solution containing (top) unpolymerized tubulin. When a voltage is applied to the membrane on left, it depolarizes. *Bottom*: after causing polymerization of tubulin (by chelating calcium), MT/tubulin bridges couple the two membranes. Depolarization of one causes depolarization of the second

Among the responses of these kinases are phosphorylation/dephosphorylation of MAPs and intermediate (neuro) filaments which can reconfigure intra-neuronal structural architecture.

Some evidence circumstantially links the cytoskeleton with cognitive function. For example, Mileusnic et al. (1980) correlated production of MT subunit protein ("tubulin") and MT activities with peak learning, memory, and experience in baby chick brains. Cronley-Dillon et al. (1974) showed that when baby rats begin their critical learning phase for the visual system (when they first open their eyes), neurons in the visual cortex begin producing vast quantities of tubulin. Tubulin production is drastically reduced when the critical learning phase is over (when the rats are 35 days old). Moshkov et al. (1992) showed structural and chemical changes in goldfish brain neuronal cytoskeleton following sensory stimulation. In gerbils exposed to insufficient brain blood flow, Kudo et al. (1990) correlated the amount of reduction in dendritic MAP2 with the degree of cognitive impairment. Kwei et al. (1993) showed that oxygen lack caused loss of MAP2 in rat hippocampus and Taft et al. (1993) showed that loss of hippocampal MAP2 following traumatic brain injury in rats could be attenuated by hypothermia which also preserved neural function. Geerts et al. (1992) showed that sabeluzole, a memory-enhancing drug, increases MT based fast axoplasmic transport. Bensimon and Chernat (1991) found that selective destruction of brain MT by the drug colchicine caused cognitive defects in learning and memory which mimic the clinical symptoms of Alzheimer's disease, in which the cytoskeleton becomes entangled. Matsuyama and Jarvik (1989) have proposed that Alzheimer's is a disease of MT and MAPs. In specific hippocampal regions of the brains of schizophrenic patients, Arnold et al. (1991) found distorted neuronal architecture due to a lack of 2 MAPs (MAP2 and MAP5). Kerwin (1993) has proposed that schizophrenia occurs due to defects in glutamate receptor regulation of these MAPs. Lew (1992) has shown changes in the neuronal MAP tau induced by the psychoactive drug phencyclidine.

Models of learning in mammalian NMDA (*N*-methyl-*D*-aspartate) hippocampal neurons (i. e., long term potentiation: "LTP") involve pre- and post-synaptic enhancement of synaptic function. Silva et al. (1992) have shown calcium-calmodulin kinase to be essential for LTP. Aszodi et al. (1991) have demonstrated protein kinase A to be involved in learning. Halpain and Greengard (1990) have shown that post-synaptic activation of NMDA receptors induces rapid dephosphorylation of dendrite-specific MAP2. Aoki and Siekevitz (1988) linked learning to protein kinase C mediated MAP2 phosphorylation/dephosphorylation, and Theurkauf and Vallee (1983) demonstrated that MAP2 phosphorylation/dephosphorylation consumed a huge proportion of brain biochemical energy. Bigot and Hunt (1990) showed that NMDA and other excitatory amino acid neurotransmitters caused redistribution of intraneuronal MAPs, and Wang and Rasenick (1991) have shown that G proteins directly link with MT. Surridge and Burns (1992) demonstrated that the lipid second messenger phosphatidylinositol binds specifically to MAP2, a coupling which

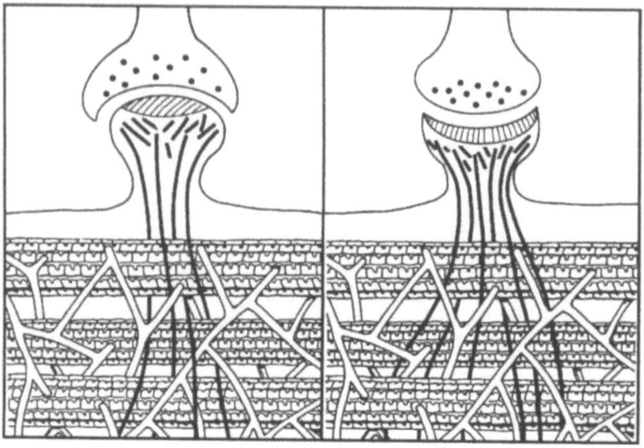

Fig. 5.6. Schematic representation of learning-associated change in dendritic spine shape due to reorganization of actin and MT cytoskeleton (Desmond and Levy, 1988). *Left*: "unlearned", *Right*: Learned (dense cytoskeleton, more receptors, lower threshold)

may "constitute a link by which extracellular factors influence the microtubule cytoskeleton." Kwak and Matus (1988) showed that denervation caused long-lasting changes in MAP distribution, and Desmond and Levy (1988) showed that LTP involved cytoskeletal-mediated shape changes in dendritic spines (Fig. 5.6). Lynch and Baudry (1987) have proposed that proteolytic digestion of the sub-synaptic cytoskeleton, followed by its structural reorganization, correlate with learning, and Friedrich (1990) has formalized a learning model of synaptic cytoskeletal restructuring. Synaptic regulation, both in learning and steady state, also depends on material (enzymes, receptors, neurotransmitters, etc.) synthesized in the cell body and transported along axons and dendrites by contractile MAPs attached to MT ("axoplasmic transport"). Similar mechanisms involving actin, synapsin, and other cytoskeletal polymers are involved in neurotransmitter release (Fig. 5.7).

Microtubules and other cytoskeletal structures are also implicated in other important cell functions. Puck and colleagues (e.g., Puck 1987; Haag et al. 1994) have argued for a number of years that the cytoskeleton regulates the genome, and that signaling along microtubules occurs as cascades of phosphorylation/dephosphorylation linked to calcium ion flux. Tsukidate et al. (1993) have shown that microtubule antagonists activate programmed cell death ("apoptosis"). Their findings are consistent

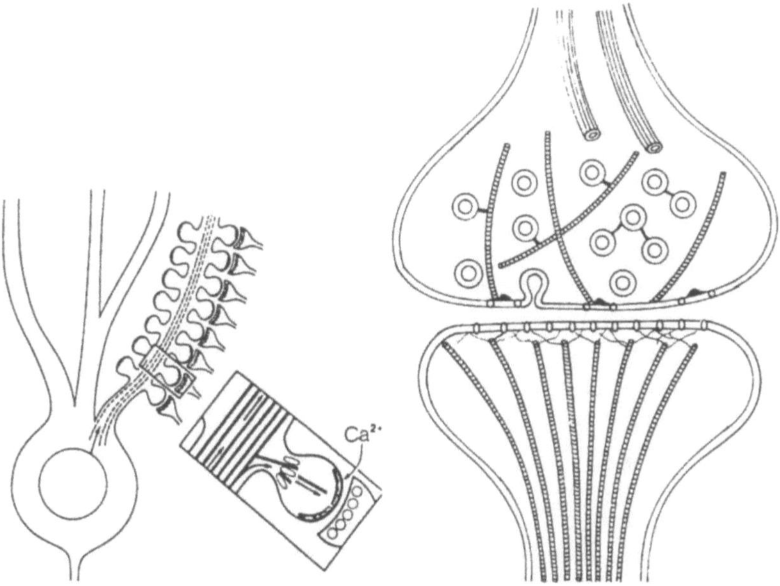

Fig. 5.7. (**a**) Schematic of neuronal dendrites and dendritic spines with internal cytoskeleton in one dendrite. *Rectangle*: close-up of dendrite and dendritic spine with cytoskeleton. Ca^{++} indicates area of calcium influx. (**b**) Schematic of cytoskeletal structures in synapse. *Top*: Pre-synaptic axon with MT, synapsin, neurotransmitter vesicles. *Bottom*: Post-synaptic dendrite with receptors, linking proteins, filaments/MT. (With permission from Hirokawa, 1991)

with observations that cytoskeletal disruption is the irreversible step in cell death from a variety of injuries. Tan et al. (1993) showed that brain MAP2 phosphorylation is decreased following prenatal alcohol exposure in rats, implying that fetal alcohol syndrome, affecting brains of children whose mothers overindulged in alcohol during pregnancy, is a disease of the cytoskeleton. The cytoskeleton provides a "missing link," a communication network among membrane, second messenger, and nuclear genetic elements (in neurons this may traverse a rather large distance). How might MT and other cytoskeletal elements signal and process information?

Tubulin subunits in closely arrayed MT have a density of about 10^{17} per cm^3, which is very close to the theoretical limit for charge separation (Gutmann 1986). Thus cytoskeletal polymers have maximal density for information representation by charge, and the capacity for dynamically coupling that information to mechanical and chemical events via quantum related co-operative conformational states.

5.4 Protein Conformational Dynamics

Proteins are vibrant, dynamic structures in physiological conditions. A variety of techniques have shown that proteins (whose sizes are typically in the range from 5 to 10 nanometers) and their component parts undergo conformational motions over a wide range of time scales from 10^{-15} seconds to many minutes, and energy scales from 0.1 to 100 kilocalories per mole (Karplus & McCammon, 1983). Very rapid (10^{-15} sec) changes occur in protein side chains or local regions, whereas slow changes may be due to binding of ligands such as oxygen to hemoglobin, and extremely long lasting changes can be induced by post-translational modifications (enzymatic addition or removal of amino acids) such as glycosylation or detyrosination. Collective conformational transitions in which proteins move globally occur in the nanosecond (10^{-9} sec) to 10 picosecond (10^{-11} sec) time scale. Related to co-operative movements of smaller regions and charge redistributions such as dipole oscillations, these global changes are linked to protein function (signal transduction, ion channel opening, enzyme action, etc.) and may be triggered by factors including phosphorylation, ATP or GTP hydrolysis, ion fluxes, electric fields, ligand binding, and "allosteric" influences by neighboring protein conformational changes.

A quantum-mechanical approach was taken by Fröhlich (1968; 1970; 1975; 1986) who proposed that protein conformational changes were triggered by quantum-level activities such as electron movements or dipole oscillations within specific hydrophobic regions of proteins (Fig. 5.8). Fröhlich (1970) further proposed that a set of proteins (and associated water molecules) connected in a common physical structure and electromagnetic field such as within a polarized membrane (or polymer electret like a microtubule) may be excited coherently if biochemical energy such as protein phosphorylation or ATP or GTP hydrolysis were supplied. Thermal dissipation of Fröhlich energy is thought to be prevented by ordered water at protein surfaces (Del Giudice et al. 1986). Coherent excitation frequencies on the order of 10^9 to 10^{11} Hz (identical to the time domain for functional protein conformational changes, and in the microwave or gigaHz spectral region) were deduced by Fröhlich who termed

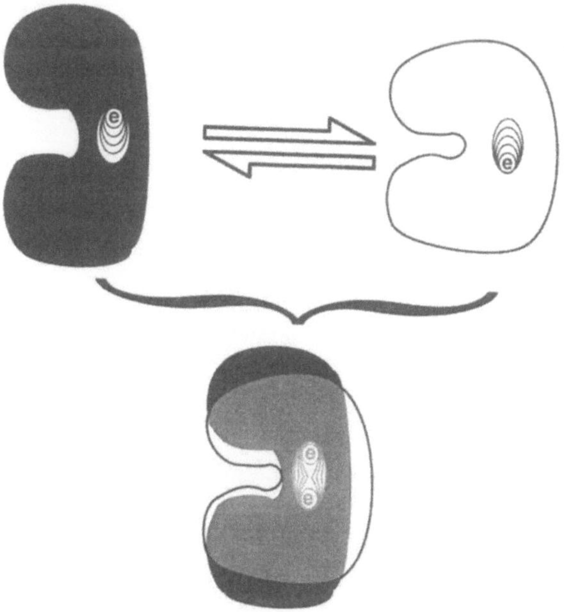

Fig. 5.8. *Top*: States of protein (e.g. tubulin) in which quantum event within hydrophobic region couples to conformational change with 10^{-9} to 10^{-11} sec transitions. *Bottom*: Quantum superposition of both states. General anesthetics bind within hydrophobic regions and prevent conformational transitions by van der Waals forces acting on quantum event

them acousto-conformational transitions, or coherent phonons. Such coherent states, examples of which include lasers, superconductors, and superfluids, are termed Bose–Einstein condensates in quantum physics. Other implications of Fröhlich's model include biological superconductivity, metastable states (longer-lived conformational state patterns stabilized by local factors), long range (non-local) communication, and polarization waves (traveling regions of dipole coupled conformations out of phase with the majority of coherently excited states).

Experimental evidence for Fröhlich-like coherent excitations in biological systems includes observation of gigaHz-range phonons in proteins (Genberg et al. 1991), sharp-resonant non-thermal effects of microwave irradiation on living cells (Grundler & Keilman 1983), gigaHz induced activation of microtubule pinocytosis in rat brain (Neubauer et al. 1990), and laser Raman spectroscopy detection of Fröhlich frequency energy (Genzel et al. 1983; Vos et al. 1992).

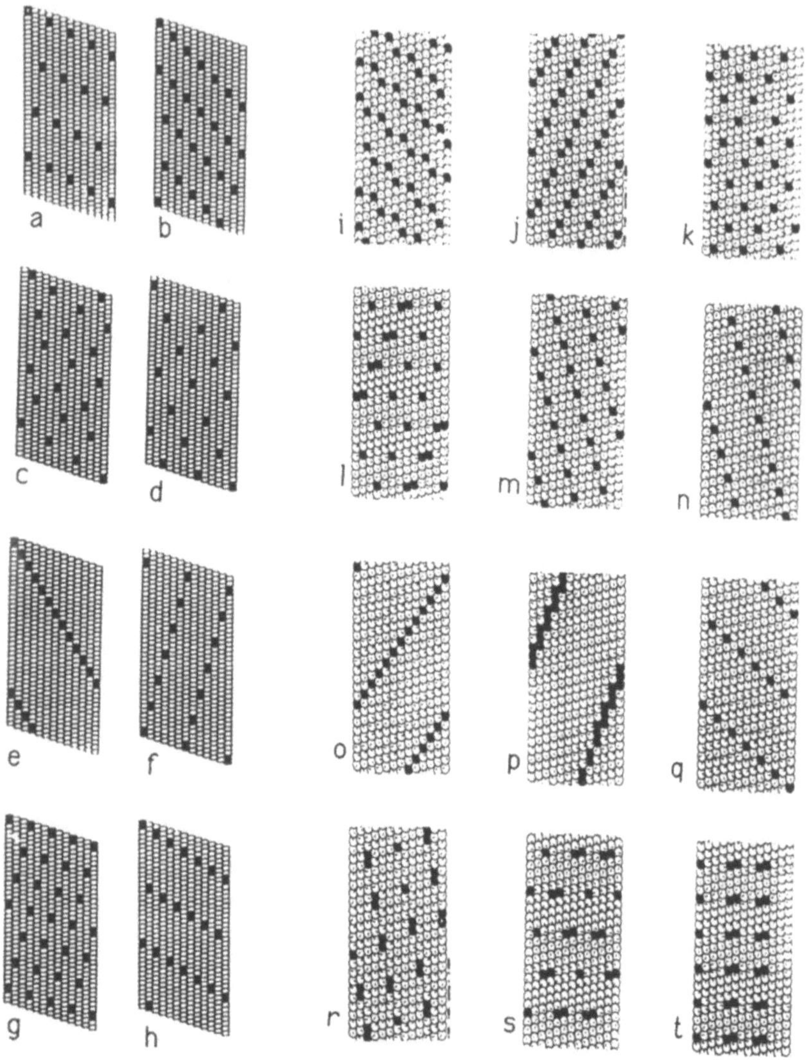

Fig. 5.9a–t. **(a)**–**(h)** Experimentally observed patterns of MAP attachment sites observed on MTs (Burns, 1978; Kim et al., 1986). **(i)**–**(t)** Patterns of MT phonon modes calculated in Samsonovich et al. (1992). Some observed patterns (**(e)** and **(f)**) exactly match and others resemble simulated phonon maxima

Attachment patterns of microtubule-associated proteins (MAPs) on microtubules may be periodic (Burns 1978; Kim et al. 1986). Because MAP attachment sites determine neuronal and synaptic function and architecture, these self-organized patterns may be examples of emergent, cognitive states. Samsonovich et al. (1992) have shown that experimentally observed patterns of MAP attachment sites on microtubules can be simulated and possibly derive from self-localized coherent phonon excitations (Fig. 5.9).

Studies on the mechanisms of general anesthetic gas molecules, which reversibly ablate consciousness while sparing other brain functions, have shown that protein conformational dynamics are the critical inhibited functions (e. g., Franks & Lieb 1982). The anesthetics occupy hydrophobic regions of proteins and prevent their conformational responsiveness to appropriate stimuli. Thus, quantum coupled conformational states of neuronal proteins appear essential to consciousness.

5.5 Models of Information Processing in Microtubules

Activities, functions and structures of microtubules and other cytoskeletal components have led at least a dozen author groups to consider theoretical models of MT based information processing (Hameroff 1987). Some are purely "classical", others consider both classical and quantum effects, and still others focus on quantum effects.

Classical Models of MT Information Processing

Atema (1973) suggested that propagated conformational changes in tubulin subunits transduced information within cilia. Hameroff and Watt (1982) proposed that these conformational changes could "switch and interact" in a computing mode within MTs. Barnett (1987) suggested that MTs act as processing elements, neurofilaments provide memory storage, and that "string processor" exchanges occur through MAP connections (Fig. 5.10). Assuming five possible conformational states of tubulin, Roth and Pihlaja (1977) theorized that co-operative allosteric effects among MT tubulin subunits created patterns of conformational gradients ("gradions") induced by MAP binding (Fig. 5.11).

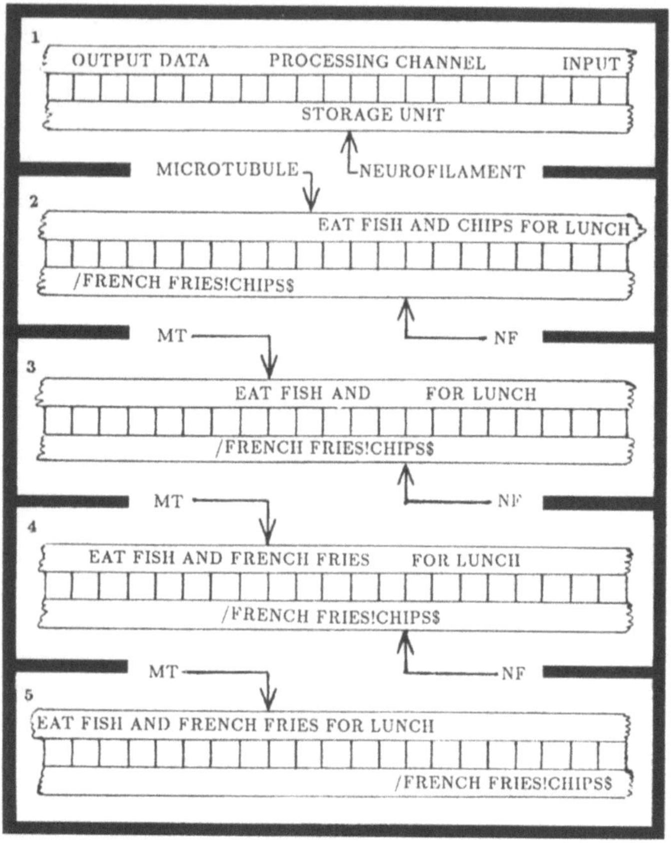

Fig. 5.10. Barnett's (1987) string processor model of microtubule-neurofilament memory storage. 1) a processing channel (microtubule-MT) is laterally connected to a parallel storage unit (neurofilament-NF), 2) information traverses the processing channel from right to left; on the storage channel, synonym pairs move from left to right, 3) as the word "chips" moves by its own self in storage, it is deleted, 4) "chips" is replaced by its synonym "french fries" and 5) the replacement is complete. By Paul Jablonka

Classical/Quantum Models of MT Information Processing

Consideration of MTs as lattice arrays of coupled (tubulin) dipoles which co-operatively interact with their immediate neighbors leads to two types of systems which can have both classical and quantum modes of information processing: spin glasses and cellular automata.

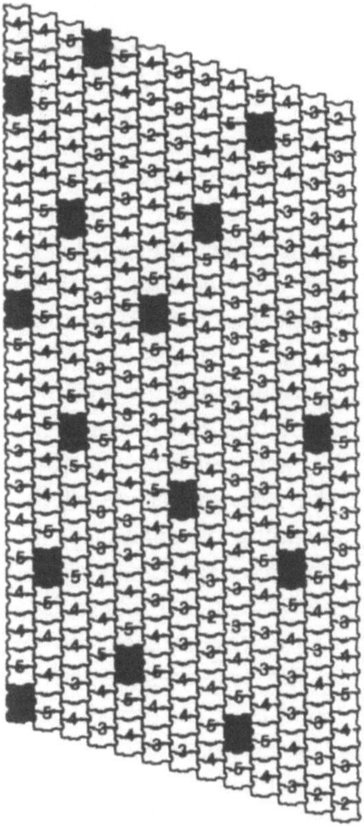

Fig. 5.11. "Gradions" (conformational gradient patterns) within microtubule lattice. Dark shaded dimers are MAP attachment patterns; numbered dimers are determined by proximity to MAP attachment sites. MAP binding and other "gradionators" are thought to induce coded patterns representing information. From Roth and Pihlaja (1977) by Paul Jablonka

In general, three types of arrangements of dipoles in lattices may occur: a) random, b) ferroelectric (parallel-aligned), and c) spin-glass (regions of locally frozen orientations). Tuszynski et al. (1995) have examined behavior of MTs in this context; depending on the values of assumed parameters, MT dipole lattices may exhibit "frustration," conflict in satisfying all dipole couplings. This phase structure is defined as a spin-glass which exists close to the edge of chaos and has properties suitable for efficient information processing and computation. By slightly altering temperature and external field (both within physiological conditions), MT dipole coupling lattices may assume a ferroelectric phase with long-range order and alignment with capabilities to propagate kink-like excitations or solitons. The ferroelectric phase appears to be optimal for propagation of signals and regulation of assembly/disassembly.

General descriptions of systems (of which computers are special cases) in which collective activities of local subunit interactions can lead to complex information processing are called "cellular automata". ("Cellular" was originally meant to connote the mistaken notion of cells in the biological sense as indivisible subunits.) Von Neumann's (1966) original cellular automaton consisted of a large number of identical "cells" connected in a uniform pattern. The essential features of cellular automata are as follows: 1) at a given time, each cell is in one of a finite number of states (usually two for simplicity). 2) The cells are organized according to a fixed geometry. 3) Each cell communicates only with other cells in its neighborhood; the size and shape of the neighborhood are the same for all cells. Depending on geometry, the number of neighbors may be 4 (rectangular), 6 (hexagonal), 8 (rectangular with corners), or more neighbors per subunit or cell. 4) There is a universal clock. Each cell may change to a new state at each tick of the clock depending on its present state and the present states of its neighbors. The rules for changing state are called the transition rules of the cellular automata. At each clock tick (or "generation") the behavior of each cell depends only on the states of its neighbors and its own state. Simple neighbor transition rules can lead to complex, dynamic patterns. For example cellular automata patterns can exhibit chaos, fractal dimensions, partial differential equations, and computation. Patterns that move through the lattice unchanged are called "gliders"; Von Neumann proved mathematically that gliders traveling through a sufficiently large cellular automaton can solve virtually any problem. Figure 5.12 shows examples of Fredkin's self-replicating cellular automata.

Wolfram (1984) observed that behaviors of cellular automata fell into a finite number of categories ("attractors") including chaos and self-organizing patterns. Wolfram further noted that "it is common in nature to find systems whose complexity is generated by the co-operative effect of many simple identical components." Wuensche (1992) has catalogued basins of attraction for one-dimensional cellular automata. Conrad (1974) used the concept of "molecular" automata within neurons (in the context of cyclic AMP fluxes) as an information processing system subserving synaptic connectionism. Grössing (1993) has described quantum

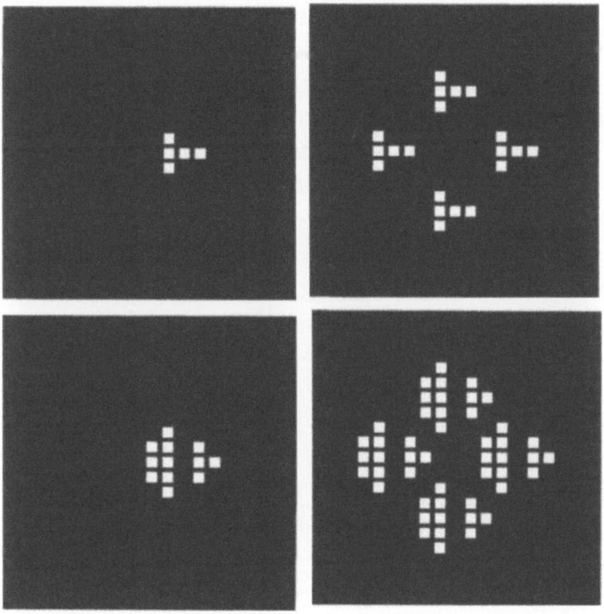

Fig. 5.12. Self-replicating automata described by Edward Fredkin

cellular automata with probabilistic states and neighbor interactions. Using Fröhlich's coherent excitations as a clocking mechanism, cellular automata principles have been applied to the dynamic conformational states of tubulin within cytoskeletal MT.

"Cellular" (molecular) automata require a lattice whose subunits can exist in two or more states at discrete time steps, and transition rules which determine those states among lattice neighbor subunits. Tubulin conformations within MT lattices can provide such states, and neighbor-tubulin interactions (represented by dipole coupling forces) may provide appropriate transition rules. A rough estimate for the time steps, assuming one coherent "sound" wave across the MT diameter ($\Phi \sim 25$ nm) and $V_{\text{sound}} = 10^3\ \text{m}\,\text{s}^{-1}$, yields a clocking frequency of approximately 4×10^{10} Hz, and a time step of 2.5×10^{-11} s. Thus Fröhlich's coherent excitations can provide a "clocking frequency" for MT conformational automata (Hameroff et al. 1989; Rasmussen et al. 1990).

Conformation of individual tubulin subunits at any given time depends on "programming" factors including initial conformational state, primary genetic structure, binding of water, ions, or MAPs, bridges to other MT, post-translational modifications, phosphorylation state, and mechanical and electrostatic dipole forces among neighboring subunits. Only electrostatic dipole forces among neighboring tubulin subunits are considered in this simulation. Figure 5.13 shows a 7-member MT automata neighborhood: a central dimer surrounded by a tilted hexagon of 6 neighbor dimers. The two monomers of each dimer share a mobile electron which is oriented either more toward the α-monomer ("alpha state") or more toward the β-monomer ("beta state") with associated changes in dimer conformation at each time step (e. g., as in Fig. 5.8). The net electrostatic force from the six surrounding neighbors acting on a central dimer can then be calculated as:

$$f_{\text{net}} = \frac{e^2}{4\pi\varepsilon} \sum_{i=1}^{6} \frac{Y_i}{r_i^3}$$

where Y_i and r_i are defined as illustrated in Fig. 5.13, e is the electron charge, and ε is the average protein permittivity. Neighbor electrostatic dipole coupling forces (on the order of 2.3×10^{-14} Newtons) are given in Hameroff et al. 1989 and Rasmussen et al. 1990.

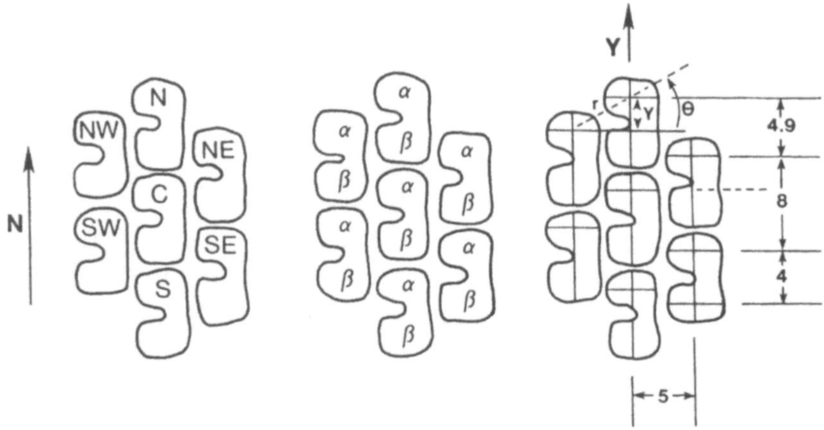

Fig. 5.13. Microtubule automata (MTA) neighborhood: left, definition of neighborhood dimers; center, a and b monomers within each dimer; right, distances in nm and orientation among lattice neighbors. Transition rules for MTA are based on net electrostatic coupling forces on each tubulin from its surrounding six tubulin neighbors. Coupling force values (described in text) are given in Hameroff et al. 1989 and Rasmussen et al. 1990

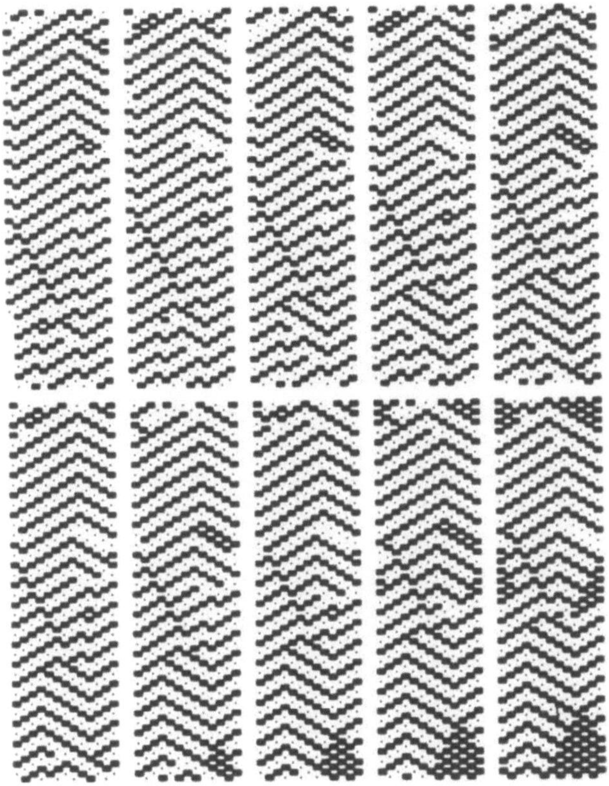

Fig. 5.14. 6 microtubule automaton simulations. Ten time steps (top-left to right, then bottom left to right) are shown; each MT depicted as vertical, flattened rectangle with neighboring edges and ends (torus). *Black*: beta state tubulin; *white*: alpha state tubulin. Information patterns evolve, move, interact. Certain virtual objects are evident to close scrutiny: kinks which travel in opposite directions, spider gliders, oscillators, etc.

To simulate MT automata, cylindrical MT structure is displayed as a 2-dimensional rectangle whose edges (adjacent protofilaments) are contiguous. To avoid boundary conditions, end borders also communicate so that a torus is modeled. MT subunit dimer loci are in either α state (white) or β state (black). At each "nanosecond" time step, forces exerted by 6 surrounding neighbor subunits are calculated for each dimer. If the net force exceeds a threshold, a transition ($\alpha \rightarrow \beta$, $\beta \rightarrow \alpha$) occurs. For example, a threshold of ± 9.0 means that net neighbor forces greater than $+9.0 \times 2.3 \times 10^{-14}$ Newtons will induce an α state, and negative forces

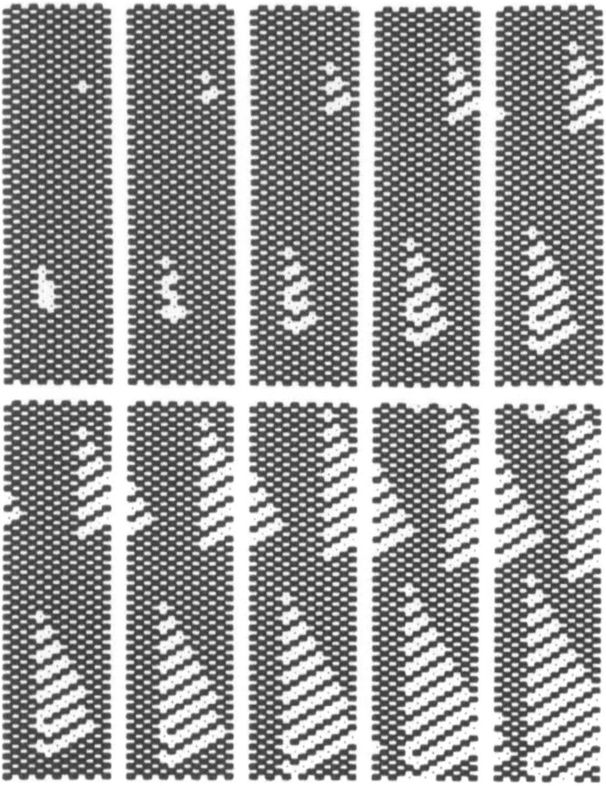

Fig. 5.15. MT automata continued

of less than $-9.0 \times 2.3 \times 10^{-14}$ Newtons will induce a β state. Threshold may represent temperature, pH, voltage gradients, ionic concentration, genetically determined variability in individual dimers, binding of molecules including MAPs and/or drugs to dimer subunits, etc. Figures 5.14–18 show MT automata simulations whose behaviors include gliders, traveling and standing wave patterns, oscillators, linearly growing patterns, and frozen patterns (perhaps suitable for memory). Asymmetrical thresholds ($\alpha \rightarrow \beta \neq \beta \rightarrow \alpha$) result in bi-directional gliders. MT automata networks (MT interconnected by MAPs) have been shown capable of learning, recognition, and back-error propagation as in artificial neural networks (Rasmussen et al. 1990; Hameroff et al. 1992). Figure 5.19 illustrates a quantum computing (Orch OR) version (see Addendum).

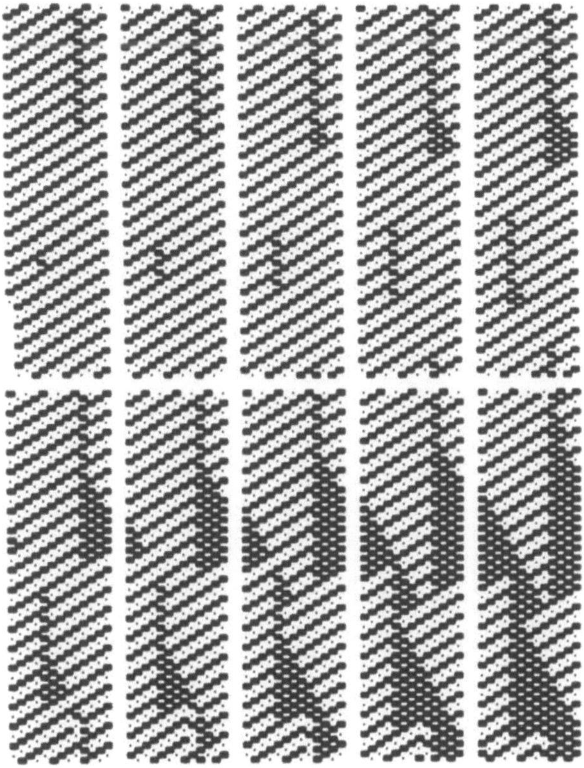

Fig. 5.16. MT automata continued

Assuming MT automata gliders and patterns exist, what functions could they have? They could represent information being signaled through the cell. Glider numerical quantities and patterns may manifest signals, binding sites for ligands, MAPs or material to be transported. Frozen patterns may store information in a memory context, information may become "hardened" in MT by post-translational modifications, and/or MT automata patterns could transfer and retrieve stored information to and from neurofilaments via MAPs.

MT automata gliders travel one dimer length (8 nm) per time step (10^{-9} to 10^{-11} s) for a velocity range of 8–800 m s^{-1} consistent with propagating solitons or phonons as well as nerve membrane potentials. Thus traveling MT automata patterns or gliders (equivalent to solitons, phonons, or Fröhlich depolarization waves) may propagate in the cy-

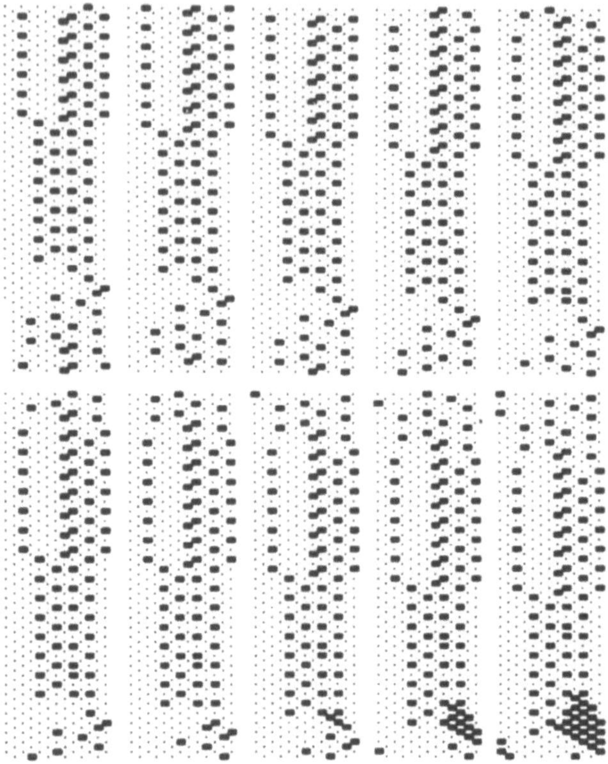

Fig. 5.17. MT automata continued

toskeleton in concert with membrane depolarizations and ion fluxes. Long range co-operativity, resonances and phase transitions among spatially arrayed MT automata patterns may occur, and membrane related voltages, ion fluxes, or direct links could induce transient waves of conformational switching or lowered threshold along parallel arrayed MT. Consequently, inputs and activities of a particular cell could directly elaborate patterns within that cell's MT automata.

Purely quantum models of MT information processing are considered in the next section.

Fig. 5.18. Microtubule automaton simulation (from Hameroff & Penrose 1996) as in Figs. 5.14–17. Eight nanosecond time steps of a segment of one microtubule are shown in "classical computing" mode in which patterns move, evolve, interact and lead to emergence of new patterns

5.6 Quantum Effects and Consciousness

Quantum theory describes fundamental properties of matter and energy which comprise our universe. Major features of quantum theory include wave/particle duality, the (Heisenberg) "uncertainty principle", superposition, randomness or acausality, and quantum inseparability or nonlocality (quantum coherence). These strange properties have been repeatedly verified by experimentation over much of this century and may help explain inexplicable phenomena related to consciousness including self, non-algorithmic processing ("insight"), free will, and the nature of human experience.

At the base of quantum theory is the wave/particle duality of atoms and their components. The Heisenberg uncertainty principle dictates that either the position or momentum of a particle (but not both) can be known with arbitrary precision. Quantum superposition implies that an unobserved atom or particle exists in many possible states; when observed, one particular state occurs, but the "choice" of states is apparently random. Generally, only very small systems are considered to be in quantum superposition; however, quantum theory as presently formulated does not exclude macroscopic quantum superposition. Thus "Schrödinger's cat" (a famous thought experiment in which release of a poison into the cat's enclosed box depends on a single quantum event) may, according to quantum superposition, be both alive and dead until observed!

However Penrose (1989; 1994; 1996) "objective reduction" proposes that an isolated quantum superposed system will self-collapse at a specific time due to a quantum gravity threshold related to the superposed mass.

Quantum non-locality, or quantum coherence implies that objects that have once interacted are in some sense still connected! Schrödinger observed in 1935 that when two quantum systems interact, their wave functions become "phase entangled." Consequently, when one system's wave function is collapsed, the other system's wave function, no matter how far away, instantly collapses too (Herbert 1993). The non-local connection (quantum coherence) is not only instantaneous and independent of distance but impervious to shielding. Experiments by Clauser et al. (1969) and Aspect and Grangier (1986) showed that measurement of one photon instantly affects the outcome of the measurement of a related, but distant photon. The quantum world is truly interconnected; *quantum entities are "aware" of the states of their spatially separated relatives!* These aspects of quantum theory have been used to help understand some of the difficult issues of consciousness.

Unitary Sense of Self

To explain the "unity of conscious experience," Marshall (1989) searched for a quantum mechanism that is extended in space (non-local, long range order), capable of many states, and has a "shared identity" unanalyzable into parts ("diversity in unity"). Marshall concludes these properties are unique to systems known as "Bose-Einstein condensates" (previously mentioned relevant to Fröhlich excitations and protein conformational dynamics), in which many quantum particle/waves move coherently, described by the same wave function.

Examples of known Bose–Einstein condensates include the laser, in which many photons occupy exactly the same optical state, superconductors, in which numerous linked electrons ("Cooper pairs") take on identical quantum possibilities, and superfluids like helium, in which quantum synchronized helium atoms create a friction-free fluid (Leggett 1989). Superconductors and superfluids require temperatures near absolute zero to reduce thermal oscillations and bring the particles into coherence. Lasers require high energy to drive crystals into coherent excited states. Of known systems akin to Bose–Einstein condensates, only Fröhlich's proposed coherent pumped phonon system can operate at body temperature.

Citing Fröhlich's pumped phonons (previously discussed) in unspecified "neural proteins" as a Bose–Einstein condensate, Marshall predicted that conscious phenomena (thoughts, images, etc.) occur as phase differences among coherent components like waves in superfluid helium or holograms in laser light. He reasoned that such a condensate could suffice even as only a minor facet of brain dynamics (in a superconductor only one material part in ten thousand participates in the phase change, but the macroscopic properties are dramatically altered). Cytoskeletal microtubules have suitable qualifications to act as a brain substrate for Bose–Einstein condensates related to consciousness. Fröhlich pumped phonons have already been considered in microtubules as a basis for information processing via spin-glass behavior and as a clocking mechanism for cellular automata functions. Microtubules' paracrystalline periodic lattice and cylindrical configuration provide an ideal coherent structure, and the functions of microtubules are crucial to intracellular cognition. Thus quantum coherence (e. g., Bose–Einstein condensate) in microtubules throughout the brain can manifest a "unitary sense of self".

Quantum coherence has been theoretically applied not only to membranes and cytoskeletal structures, but to water molecules ordered at their surfaces. Fröhlich theorized that energy could be stored without thermal loss in coherent, dipolar propagating proteins or in a thin layer of water and ions just beneath the cell membrane. Watterson (1987) has shown how organized, dynamic water clusters ("pixels") can represent information at membrane and cytoskeletal surfaces. Considering the layer of ordered water outside and inside microtubules, Del Giudice et al. (1983) proposed that microtubules' cylindrical structure may be understood by quantum vacuum symmetry breaking and boson self-focusing by ordered water (bosons are non-exclusionary quantum entities–photons, phonons, and electron pairs are bosons). Like the Meissner effect for superconducting media, electromagnetic energy would be confined inside filamentous regions around which the microtubule subunits gather. Del Giudice's group (1983) showed that this self-focusing should result in filamentous

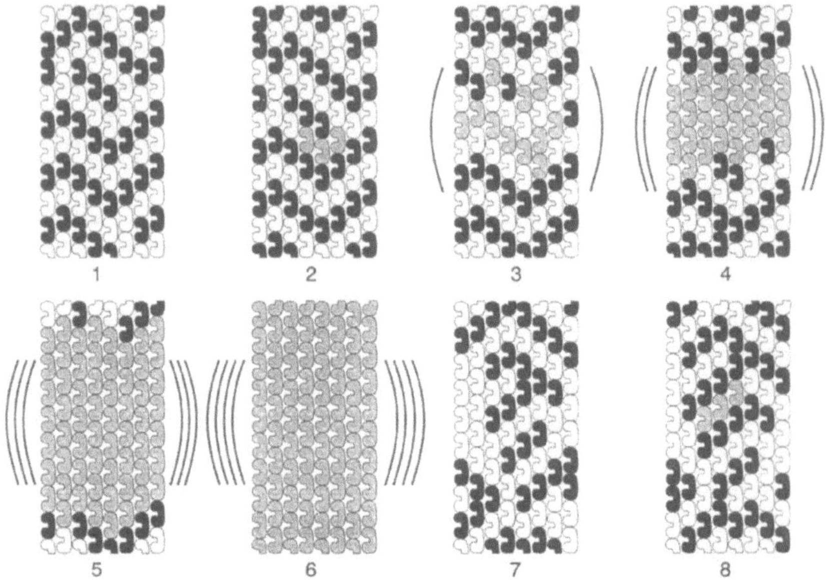

Fig. 5.19. Microtubule automaton sequence simulation in which classical computing (step 1) leads to emergence of quantum coherent superposition (steps 2–6) in certain (gray) tubulins due to pattern resonance. Step 6 (in coherence with other microtubule tubulins) meets critical threshold related to quantum gravity for self-collapse (Orch OR). Consciousness (Orch OR) occurs in the step 6 to 7 transition. Step 7 represents the eigenstate of mass distribution of the collapse which evolves by classical computing automata to regulate neural function. Quantum coherence begins to reemerge in step 8

beams of radius 15 nanometers, precisely the inner diameter of microtubules!

Using quantum field theory, Jibu et al. (1994) have proposed that ordering of water molecules and the quantized electromagnetic field confined inside the hollow microtubule core manifest a specific collective dynamics called *super-radiance*. Accordingly, each microtubule can transform incoherent, disordered energy (molecular, thermal, or electromagnetic) into coherent photons within its hollow core. Furthermore, coherent photons created by super-radiance may penetrate without loss along the microtubule as if the optical medium were made "transparent" by the propagating photons themselves. This is a quantum phenomenon called *self-induced transparency*. Some evidence suggests weak photoemission from living cells, and Popp (1986) has proposed a regulatory role for "coherent biophotons".

Quantum MT states may be isolated by cycles of actin gelation, and involve multiple neurons through gap junctions.

About 15% of brain synapses are electrical synapses, now recognized to be largely "gap junctions" membrane pores which connect with similar pores in an adjacent neuron's membrane. Cell cytoplasm is continuous among gap junction-connected neurons which fire synchronously, behaving like "one giant neuron" (Kandel et al. 1991). Gap junction separations are 3.5 nanometers, within range for quantum tunneling. Jibu (1992) proposed that gap junctions/electrical synapses account for synchronized 40 Hz neural activity, suggested to provide "binding" in classical consciousness models (e. g. Crick & Koch 1990). Networks of gap junction-connected neurons could isolate quantum coherent MTs across large brain volumes.

Insight/Non-algorithmic Processing

Penrose (1989) has argued that our brains perform computations which are "non-algorithmic" – impossible by conventional serial or parallel processing. He concludes that quantum computing taking advantage of coherence and superposition is somehow utilized.

Quantum superposition and coherence have been linked to specific biological function as well as technological schemes for computation. Conrad (1992) has argued that quantum coherence and superposition can enhance the speed, reliability, and energy dissipation of elementary switches such as those involved in molecular computation (including, perhaps, tubulins within microtubules). Theoretical descriptions of technological quantum computers propose utilizing quantum coherence, superposition, and parallelism. For example, Deutsch (1985; 1992) and Lloyd (1993) foresee quantum-level bit states (e. g., 1, 0) which can be in quantum superposition of both 1 and 0. If the bit state sites (in an array or polymer) are locally interactive as in a cellular automaton (e. g., from overlap among unit-unit electron wave functions, Van der Waals forces, conformational changes of subunits, dipole coupling, etc.), then parallel logic and quantum coherence can result in quantum computation; multiple computations merge into a macroscopic quantum state result. Penrose (1989) speculates that conscious thinking involves resolution of alternatives that were previously in linear superposition.

Coherent quantum phenomena cannot perform computation if the system undergoes significant "measurement-like interactions" with the outside world; only well isolated quantum computers are possible. Internal information within quantum computers inaccessible to external observation implies a "subjectivity" suggestive of consciousness (Deutsch 1992). But quantum computers must somehow interact with the external, classical world to be useful. "Classical" interfaces proposed for quantum computers include "selective driving of resonances" and arrays of weakly coupled quantum systems subjected to sequences of electromagnetic pulses of well defined frequency and length, permitting loading, processing, and unloading of information (Lloyd 1993). Cytoskeletal polymers subjected to neuronal firing might constitute such a system, and be isolated from external interaction/observation by cycles of actin gelation ("sol-gel" transitions).

Free Will

Connectionist and other computer-like models of brain function suffer from (among other problems) the issue of determinism. In general, for given inputs, outputs are determined. This behaviorist construct confines consciousness to the status of an epiphenomenon. To address this problem, Beck and Eccles (1992) proposed that quantum randomness results in what appears to be probabilistic, acausal synaptic function. Noting that in an apparently random fashion only one of approximately six axonal action potentials results in synaptic release of neurotransmitter vesicles, they pinpoint the quantum effect as the "movement of a hydrogen bridge by electronic rearrangement" acting on vesicle release through the presynaptic vesicular grid. A hexagonal lattice on the inner surface of the presynaptic region, the vesicular grid was chosen by Beck and Eccles as the site for quantum influence because of its "paracrystalline structure" which "makes it possible to have long-range interactions between the constituents." (A stronger argument can be made for more ubiquitous paracrystalline microtubule arrays.) Other candidates for mediation of quantum randomness effects include ion channels and ions (e. g., calcium ions) themselves (Herbert 1993). Quantum randomness, or uncertainty, however, fails to truly address free will. Rather it is a mere "fly in the ointment" of determinism.

Free will may lie in the cause of wave function collapse. Penrose (1989; 1994) argues that superposed systems which self-collapse by quantum gravity-mediated objective reduction reflect non-computability-influence by hidden quantum mathematical logic. Free will may lie in the net effect of this hidden logic on our otherwise deterministic actions.

There appears to be a fine line between determinism and randomness, and self-induced MT wave function collapse may lie on it.

The Essential Nature of Human Experience

Chalmers (1996) describes the “hard problem” in consciousness as the essential nature of experience, our “inner life”. He observes that even if the activity of every neuron, synapse, protein and ion in a conscious brain were correlated with a given mental state, that state’s qualia, or experience, would still be unexplained. Chalmers concludes that experience is fundamental – an irreducible feature of the universe-like mass, spin and charge.

This view is in line with panpsychist/panexperiential philosophy. Leibniz (e. g. 1766) saw the universe as an infinite number of fundamental units (“monads”) – each having a primitive psychological being. Whitehead (e. g. 1929) described dynamic monads with greater spontaneity and creativity, interpreting them as mind-like entities of limited duration (“occasions of experience” – each bearing a quality akin to “feeling”). More recently Wheeler (e. g. 1990) described a “pre-geometry” of fundamental reality comprised of information. Chalmers (1996) contends that fundamental information includes “experiential aspects” leading to consciousness.

Whitehead’s panexperiential view remains most consistent with modern physics: 1) consciousness is a process of events occurring in a wider, basic field of raw proto-conscious experience 2) Whitehead’s events (discrete occasions of experience) are comparable to quantum state reductions (Shimony 1993).

The “basic field of raw, proto-conscious experience” may be the fundamental level of reality – the “Planck scale” of 10–33 cm at which spacetime is no longer smooth, and may be described, for example, by quantum spin networks (Penrose 1971; Rovelli & Smolin 1995a; 1995b). This suggests that consciousness may involve a self-organizing process of quantum state reductions occurring at the Planck scale.

5.7 Anesthesia, the Other Side of Consciousness

Brains of patients under general anesthesia are commonly quite active: EEG, evoked potentials, respiratory and autonomic drives and other brain functions persist despite a lack of consciousness. Thus, at just the right level, general anesthesia *is* the *absence* of consciousness. A variety of different types of gas molecules are general anesthetics; they range from halogenated hydrocarbons to ethers to inert gases such as xenon. What they share in common are 1) solubility in a specific "hydrophobic" environment, and 2) binding by weak van der Waals forces. After many years of investigating effects in hydrophobic *lipid* regions of membranes, anesthetic gas molecules are now recognized to reversibly inhibit consciousness by weak, van der Waals binding in hydrophobic regions of some set of neural *proteins* (Franks & Lieb 1982). Quantum events linked to consciousness (coupled to classical, protein conformational dynamics) are predicted to occur in these same hydrophobic pockets within neural proteins (cytoskeletal, receptors, channels, second messengers, enzymes, etc.). These non-polar, lipid-like, water excluding regions are precisely where general anesthetic gas molecules exert their effects by very weak, van der Waals forces (e. g., instantaneous dipole coupling forces). Anesthetics are known to retard mobility of electrons (Hameroff & Watt 1983), and by doing so in hydrophobic regions, anesthetics may prevent protein conformational responsiveness and coupling to quantum events.

If quantum coherence in microtubules is essential for consciousness, anesthetics must somehow inhibit or prevent it, either indirectly or directly. Possible indirect ways include inhibition of activities of membrane, second messenger or membrane-cytoskeletal linking proteins which could disconnect and isolate cytoskeletal dynamics from the external milieu. Anesthetics may also disrupt hydrophobic links among microtubule-associated-proteins (MAPs) which interconnect microtubules into functional networks (Lewis et al. 1989). Alternatively, anesthetics may directly alter cytoskeletal quantum coherence. General anesthetics are known to bind to microtubules and at high enough concentrations cause their depolymerization (Allison & Nunn 1968). At concentrations of anesthesia just sufficient for ablation of consciousness, anesthetic action at tubulin hydrophobic regions could have profound effects by retarding mobility of dipoles/electrons in tubulin hydrophobic regions, thereby inhibiting tubulin conformational dynamics and preventing quantum coherent superposition. Wulf and Featherstone (1957) showed that anesthetic-protein binding within hydrophobic regions alters protein-water binding at the protein surface. Thus anesthetic-tubulin interaction would alter co-operative water binding and ordering, with consequent effects on quantum vacuum symmetry breaking and attendant "superradiance" and "self-induced transparency" (Jibu et al. 1994). By having such effects in hydrophobic regions of tubulin and other proteins, anesthetics can prevent quantum coherent superposition and consciousness.

5.8 Conclusion

Emergent phenomena arising from the brain's connectionist, neural-level architecture reveals a great deal about cognitive function. However, the difficult questions (unitary sense of self, insight/non-algorithmic processing, non-deterministic free will and the essential nature of our experience) demand a deeper understanding. Quantum coherent superposition and self-collapse (objective reduction) at the level of the intra-neuronal cytoskeleton can address these questions.

Acknowledgement. Stuart Hameroff is partially supported by NSF Grant No. DMS-9114503, and is indebted to Roger Penrose, Steen Rasmussen, Alwyn Scott, and Ian Marshall for collaboration, discussion and review. The excellent technical support of Richard Hofstad is gratefully appreciated.

Addendum

The Penrose–Hameroff "Orch OR" Model

The Abisko meeting enabled a friendship and collaboration with Roger Penrose. We had met briefly in 1992 and realized that our areas of work interacted synergistically. In *The Emperor's New Mind* (and subsequently, *Shadows of the Mind*) Penrose (1989; 1994) reasoned that consciousness involved a process of "self-collapse" of the quantum wave function by an objective criterion related to quantum gravity: objective reduction (OR). He concluded convincingly that such a process somehow occurred in the brain. I suggested that microtubules were perhaps the most suitable possible substrate for biological OR.

At Abisko we began to discuss these possibilities further. On a memorable walk along the shore of frozen Lake Torne traesk, the seeds of a model germinated. We have conferred on a hike in the Grand Canyon, a Copenhagen summer and a houseboat on Lake Powell. Over the course of these discussions we developed specific proposals for quantum coherent superposition in microtubules, the quantum gravity threshold for conscious events, biological orchestration by microtubule-associated proteins, quantum isolation and macroscopic states. Our first collaborative publication on "orchestrated objective reduction" (Orch OR) in microtubules appeared in 1995 (Penrose & Hameroff 1995) and another followed in 1996 (Hameroff & Penrose 1996a). To address the "hard problem" of the nature of conscious experience (e. g. Chalmers 1996), Orch OR embraced a panpsychist/panexperiential philosophical stance connected to reality at the Planck scale of spacetime geometry (Hameroff & Penrose 1996b). [Additional relevant references include Penrose (1997), and Hameroff (1996; 1997a; 1997b; 1997c).]

Spanning philosophy, physics and biology, the Orch OR model suggests an intimate link between consciousness and the universe.

Summary of the Penrose–Hameroff Orch OR Model

Here the basic elements of the Penrose–Hameroff Orch OR model are summarized in 11 points:

1) Conformational states of tubulins in neuronal MTs are coupled to internal quantum events (Fig. 5.8), and cooperatively interact with other MT tubulins in both classical and quantum computation (Figs. 5.14–19; Hameroff et al. 1992; Rasmussen et al. 1990).

2) Quantum coherent superposition occurs among MT tubulins, pumped by thermal and biochemical energy in a manner described by Fröhlich (1968; 1975). Actin gel states surround and isolate MT quantum coherence cyclically, in phases of sol-gel transitions (Fig. 5.20).

3) Quantum coherence in MTs and actin gelation in individual neuronal and glial cells are linked by gap junction-mediated synchronization (e. g. 40 Hz – Jibu 1990) providing macroscopic quantum coherence in networks of many gap junction-connected cells (neurons and glia) throughout large brain volumes (Fig. 5.21).

4) Quantum coherent superposition/computation in neural (and glial) MTs (quantum automata) corresponds to pre-conscious processing, which continues until the mass-distribution differences among the separated tubulins reaches a threshold, and self-collapse (objective reduction) then occurs (Fig. 5.20). These events are conscious events (see point 11).

The critical degree of spacetime separation causing Penrose's objective reduction is related to quantum gravity and given by the uncertainty principle $E = \hbar/T$. E is the gravitational self-energy of the superposed mass (displaced from itself by the diameter of its atomic nuclei), $\hbar$ is Planck's constant over $2B$, and T is the coherence time until self-collapse.

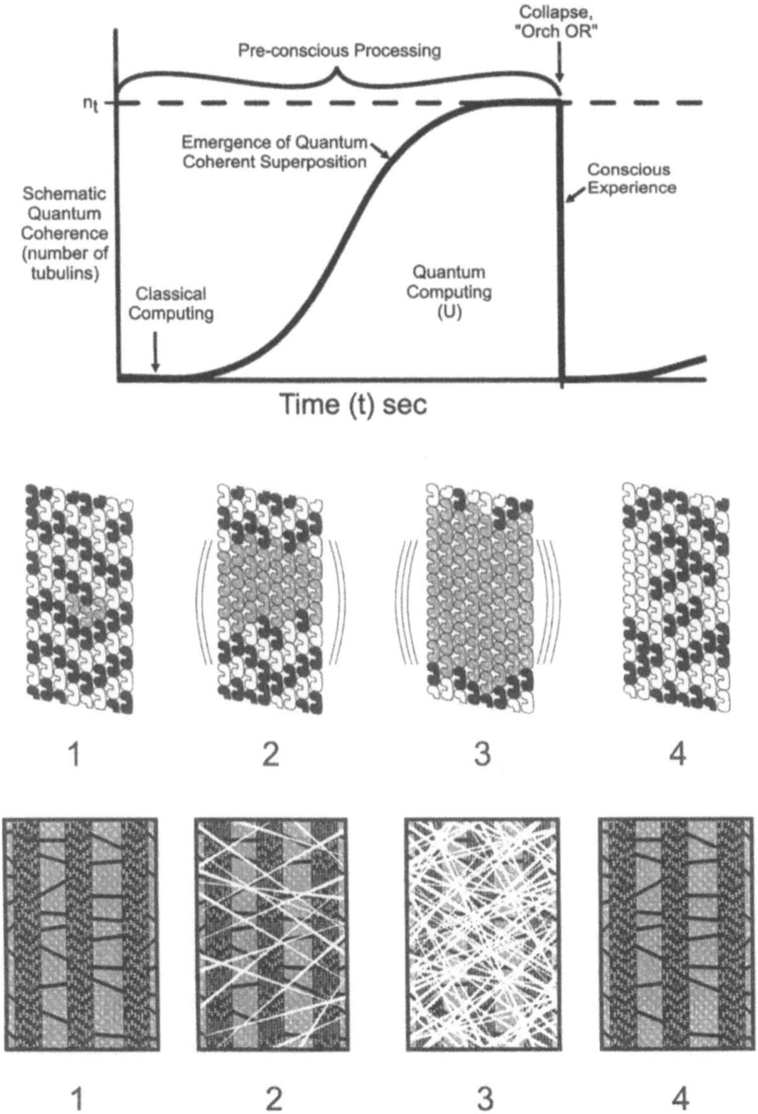

Fig. 5.20. Steps in an Orch OR event. *Top*: Schematic of proposed quantum coherence (number of tubulins) emerging in microtubules. *Middle*: Microtubule simulation in which classical computing (step 1) leads to quantum coherent superposition (and quantum computing – steps 2–3) in certain (gray) tubulins. Critical superposition (quantum gravity threshold) occurs in the step 3 to 4 transition. Tubulin states in step 4 are non-computably chosen in the objective reduction (self-collapse) and evolve by classical computing to regulate neural function. *Bottom*: Schematic sequence of phases of actin gelation (quantum isolation) and solution (environmental communication) around MTs

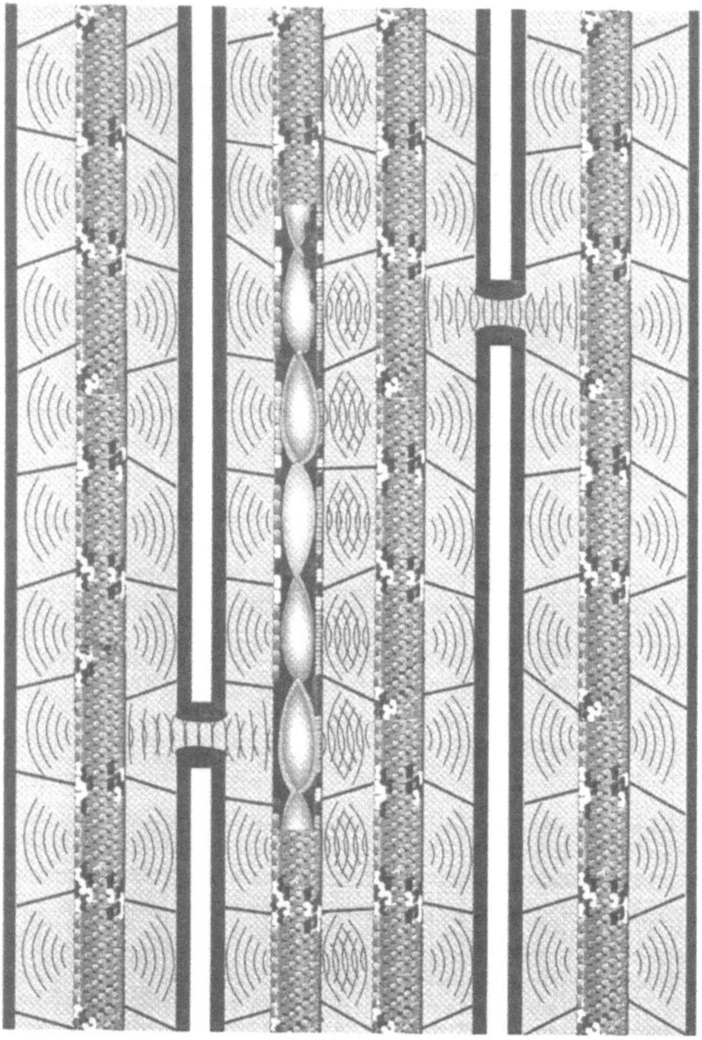

Fig. 5.21. Schematic quantum coherence in microtubules in three synchronized dendrites interconnected by tunneling through gap junctions. Within each neuronal dendrite, MAP (microtubule-associated-protein) attachments breach isolation and prevent quantum coherence; MAP attachment sites thus act as "nodes" which tune and orchestrate quantum oscillations and set possibilities and probabilities for collapse outcomes ("orchestrated objective reduction: Orch OR"). Gap junctions enable quantum tunneling among, e. g., dendrites in macroscopic quantum coherent state

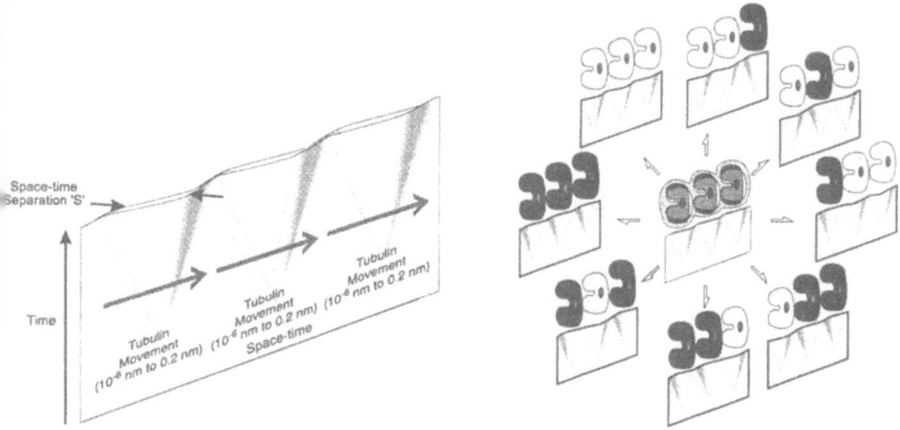

Fig. 5.22. Spacetime representations of tubulin superpositions. *Left*: Schematic space-time separation illustration of three superposed tubulins. The Planck-scale spacetime separations S are very tiny in ordinary terms, but relatively large mass movements (e. g. hundreds of tubulin conformations, each moving from 10^{-6} nm to 0.2 nm) indeed have precisely such very tiny effects on the space-time curvature. A critical degree of separation causes an abrupt selection of one curvature, or the other. *Right*: *Center* – Three superposed tubulins with corresponding schematic space-time separation illustrations. Surrounding the superposed tubulins are the eight possible post-reduction "eigenstates" for tubulin conformation, and corresponding space-time geometry. The post-reduction state is chosen noncomputably in Orch OR events

(Without an objective criterion for reduction, spacetime separation could presumably continue and result in separate, multiple spacetime universes as described in the Everett "multi-worlds" or "multi-minds" view of quantum theory.)

If isolated from environmental decoherence, a quantum superposed mass E will self-collapse after time T to definite mass locations and spacetime geometry. The post-reduction mass locations and spacetime geometry are chosen non-computably – but only if the quantum collapse occurs by the OR process, rather than by environmental interaction causing decoherence (in which case the states are chosen randomly). The non-computable influence in OR may be a Platonic grain in Planck-scale geometry. 5) Each OR self-collapse selects (non-computably) particular outcome states of MT tubulins which implement neural functions including synaptic regulation. Particular experience in funda–mental spacetime geometry is also selected (Fig. 5.22).

6) Probabilities and possibilities for post-OR states are influenced by factors including attachments of MT-associated proteins ("MAPs"), which tune and orchestrate quantum oscillations (Fig. 5.21). We thus term the self-tuning OR process in MTs "orchestrated" objective reduction – Orch OR.

7) Orch OR conscious events may be of variable intensity and duration. Calculating from $E = \hbar/T$, for a pre-conscious processing time of, for example, $T = 500$ msec (e. g. shown by Libet et al. 1979, to be one possible pre-conscious processing time), E is equivalent to 10^9 tubulins. Thus 10^9 tubulins in isolated quantum coherence for 500 msec will self-collapse (Orch OR) and constitute a conscious event (Whitehead "occasion of experience"). As previously mentioned, for pre-conscious times T of, say, 25 msec (e. g. 40 Hz neural excitations) 2×10^{10} quantum superposed tubulins would be required. A more intense 5 msec event would involve 10^{11} tubulins.

8) Each brain neuron is estimated to contain about 10^7 tubulins (Yu & Bass 1994). If, say, 10% of each neuron's tubulins became coherent, then tubulins within roughly 1,000 neurons would be required for a 500 msec Orch OR, or 20,000 for a 25 msec (40 Hz) Orch OR.

9) Each instantaneous Orch OR event can bind various superpositions which have evolved in separated spatial distributions and over differing time scales, but whose net displacement reaches threshold at a particular moment: information is bound into a conscious now. Cascades of Orch ORs can then represent our familiar stream of consciousness.

10) According to the arguments for OR put forth in Penrose (1994), superpositioned states each have their own spacetime geometry. When the degree of coherent mass-energy difference leads to sufficient separation of spacetime geometry, the system must choose and decay (reduce, collapse) to a single universe state. Thus Orch OR involves self-selections in fundamental space-time geometry.

11) A panexperiential view has evolved from Spinoza and Leibniz to Whitehead, Wheeler and Chalmers, and concludes that qualia, or experience is fundamental to the universe. We suggest that experience is therefore embedded in spacetime geometry. The Penrose spin network model of Planck scale volume quantization (Rovelli & Smolin 1995a; 1995b) is suggested here to be a possible makeup of physical reality suitable for containing qualia and raw experience ("quantum monads"). Each Orch OR conscious event is a self-configuration of funda–mental geometry.

Conclusion

The Orch OR model is consistent with known neural and cognitive activities and addresses enigmatic features of consciousness (e. g. nature of experience, binding, pre-conscious to conscious transition, flow of time, free will and non-computability). The model attempts to link brain functions with self-configurations in "funda–mental" spacetime geometry, and suggests an intimate link between consciousness and the universe.

References

Allison, A. C. & Nunn, J. F. (1968). "Effects of general anaesthetics on microtubules. A possible mechanism of anaesthesia", *Lancet* 2: 1326–1329

Amos, L. A.& Amos, W. B. (1991). *Molecules of the cytoskeleton*, Guilford Press, New York

Amos, L. A. & Klug, A. (1974). "Arrangement of subunits in flagellar microtubules", *J. Cell Sci.* 14: 523–550

Aoki, C. & Siekevitz, P. (1988). "Plasticity in brain development", *Sci. Am.* (December), 34–42

Arnold, S. E., Lee, V. M. Y., Gur, R. E. & Trojanowski, J. Q. (1991). "Abnormal expression of two microtubule-associated proteins (MAP2 and MAP5) in specific subfields of the hippocampal formation in schizophrenia", *Proc. Natl. Acad. Sci.* 88: 10850–10854

Aspect, A. & Grangier, P. (1983). "Experiments on Einstein-Podolsky-Rosen-type correlations with pairs of visible photons" in: *Quantum Theory and Measurement* ed. by J.A. Wheeler & W.H. Zurek, Princeton University Press

Aszodi, A., Muller, V., Friedrich, P. & Spatz, H. C. (1991). "Signal convergence on protein kinase A as a molecular correlate of learning", *Proc. Natl. Acad. Sci.* 88: 5832–5836

Atema, J. (1973). "Microtubule theory of sensory transduction", *J. Theor. Biol.* 38: 181–190

Athenstaedt, H. (1974). "Pyroelectric and piezoelectric properties of vertebrates", *Ann. NY Acad. Sci.* 238: 68–93

Barnett, M. P., (1987). "Molecular systems to process analog and digital data associatively" in *Proceedings of the 3rd Molecular Electronic Device Conference* ed. by F. Carter, Naval Research Laboratory, Washington, DC

Beck, F. & Eccles, J. C. (1992). "Quantum aspects of brain activity and the role of consciousness", *Proc. Natl. Acad. Sci. USA* 89(23): 11357–11361

Bensimon, G. & Chernat, R. (1991). "Microtubule disruption and cognitive defects: effect of colchicine on learning behavior in rats", *Pharmacol. Biochem. Behavior* 38: 141–145

Bigot, D. & Hunt, S. P. (1990). "Effect of excitatory amino acids on microtubule-associated proteins in cultured cortical and spinal neurons", *Neurosci. Lett.* 111: 275–280

Burns, R. B. (1978). "Spatial organization of the microtubule associated proteins of reassembled brain microtubules", *J. Ultrastruct. Res.* 65: 73–82

Burnside, B. (1974). "The form and arrangement of microtubules: an historical, primarily morphological review", *Ann. NY Acad. Sci.* 253: 14–26

Chalmers, D. J. (1996a). *The conscious mind – In search of a fundamental theory*, Oxford University Press, New York

Chalmers, D. J. (1996b). "Facing up to the problem of consciousness" in *Toward a Science of Consciousness – The First Tucson Discussions and Debates*, ed by S. R. Hameroff, A. Kaszniak and A. C. Scott, MIT Press, Cambridge, MA, pp. 5–28

Clauser, J. F., Horne, A. H., Shimony, A. & Holt, R. A. (1969). "Proposed experiment to test local hidden variable theories", *Phys. Rev. Lett.* 23: 880–884

Conrad, M. (1974) "Molecular automata" in *Lecture Notes in Biomathematics, Vol. 4: Physics and Mathematics of the Nervous System* ed. by M. Conrad, W. Guttinger & M. Dal Cin, Springer-Verlag, Heidelberg, pp. 419–430

Conrad, M. (1992). "Quantum molecular computing: the self-assembly model", *Int. J. Quantum Chemistry: Quantum Biology Symposium* 19: 125–143

Crick, F. & Koch, C. (1990). "Towards a neurobiological theory of consciousness", *Seminars in the Neurosciences* 2: 263–275

Cronly-Dillon, J., Carden, D. & Birks, C. (1974). "The possible involvement of brain microtubules in memory fixation", *J. Exp. Biol.* 61: 443–454

Dayhoff, J., Hameroff, S., Lahoz-Beltra, R. & Swenberg, C. E. (1994). "Cytoskeletal involvement in neuronal learning: A review", *Eur. Biophys. J.* (in press)

DeBrabander, M. (1982). "A model for the microtubule organizing activity of the centrosomes and kinetochores in mammalian cells", *Cell Biol. Intern. Rep.* 6: 901–915

Del Giudice, E., Doglia, S. & Milani, M. (1983). "Self-focusing and ponderomotive forces of coherent electric waves: A mechanism for cytoskeleton formation and dynamics" in *Coherent Excitations in Biological Systems* ed. by H. Fröhlich & F. Kremer, Springer-Verlag, Berlin, pp. 123–127

Del Giudice, E., Preparata, G. & Vitiello, G. (1989). "Water as a true electric dipole laser", *Phys. Rev. Lett.* 61: 1085–1088

Del Giudice, E., Doglia, S., Milani, M. & Vitiello, G. (1986). "Electromagnetic field and spontaneous symmetry breaking in biological matter", *Nucl. Phys.* B275[FS17], 185–199

Dennett, D. (1991). *Consciousness Explained*, Little Brown, New York

Desmond, N. L. & Levy, W. B. (1988). "Anatomy of associative long-term synaptic modification" in *Long-Term Potentiation: From Biophysics to Behavior* ed. by P. W. Landfield & S. A. Deadwyler

Deutsch, D. (1985). "Quantum theory, The Church-Turing principle and the universal quantum computer", *Proc. R. Soc. Land.* A 400: 97–117

Deutsch, D. (1992.). "Quantum computation", *Physics World*, 57–61
Dustin, P. (1984). *Microtubules*, 2nd Revised Ed., Springer-Verlag, Berlin
Engelborghs, Y. (1992). "Dynamic aspects of the conformational states of tubulin and microtubules", *Nanobiology* 1: 97–105
Eriksson, J. E., Brautigan, D. L., Vallee, R., Olmsted, J., Fujiki, H. & Goldman, R. D. (1992). "Cytoskeletal integrity in interphase cells requires protein phosphatase activity", *Proc. Natl. Acad. Sci.* 89: 11093–11097
Franks, N. P. & Lieb, W. R. (1982). "Molecular mechanisims of general anesthesia", *Nature* 300: 487–493
Friedrich, P. (1990). "Protein structure: the primary substrate for memory", *Neurosci.* 35: 1–7
Fröhlich, H. (1968). "Long-range coherence and energy storage in biological systems", *Int. J. Quantum Chem.* 2: 641–9
Fröhlich, H. (1970). "Long range coherence and the actions of enzymes", *Nature* 228: 1093
Fröhlich, H. (1975). "The extraordinary dielectric properties of biological materials and the action of enzymes", *Proc. Natl. Acad. Sci.* 72: 4211–4215
Fröhlich, H. (1986). "Coherent excitations in active biological systems" in *Modern Bioelectrochemistry* ed. by F. Guttmann & H. Keyzer, Plenum Press, New York, pp. 241–261
Geerts, H., Nuydens, R., Nuyens, R., Cornelissen, F., De Brabander, M., Pauwels, P., Janssen, P. A. J., Song, Y. H. & Mandelkow, E. M. (1992). "Sabeluzole, a memory-enhancing molecule, increases fast axonal transport in neuronal cell cultures", *Experimental Neurology* 117: 36–43
Genberg, L., Richard, L., McLendon, G. & Dwayne-Miller, R. J. (1991). "Direct observation of global protein motion in hemoglobin and myoglobin on picosecond time scales", *Science* 251: 1051–1054
Genzel, L., Kremer, F., Poglitsch, A. & Bechtold, G. (1983). "Relaxation processes on a picosecond time scale in hemoglobin and polylysine observed by millimeter-wave spectroscopy", *Biopolymers* 22: 1715–1729
Grössing, G. (1993). "Simulation of context dynamics with quantum cellular automata; the 'metagraph' model" (in press).
Grundler, W. & Keilmann, F. (1983). "Sharp resonances in yeast growth prove nonthermal sensitivity to microwaves", *Phys. Rev. Lett.* 51: 1214–1216
Gutmann, F. (1986). "Some aspects of charge transfer in biological systems" in *Modern Bioelectrochemistry* ed. by F. Gutmann & H. Keyzer, Plenum Press, New York, pp. 177–197
Halpain, S. & Greengard, P. (1990). "Activation of NMDA receptors induces rapid dephosphorylation of the cytoskeletal protein MAP2", *Neuron* 5: 237–246
Haag, M. M., Krystosek, A., Arenson, E. & Puck, T. T. (1994). "Reverse transformation and genome exposure in the C6 glial tumor cell line", *Cancer Investigation* 12(1): 33–45
Hameroff, S. R. (1987). *Ultimate Computing: Biomolecular Consciousness and Nanotechnology*, North-Holland, Amsterdam

Hameroff, S. R. (1994). "Quantum coherence in microtubules: a neural basis for emergent consciousness?" *Journal of Consciousness Studies* 1: 91–118

Hameroff, S. (1996). "Cytoplasmic gel states and ordered water: Possible roles in biological quantum coherence". *Proceedings of the Second Advanced Water Symposium*, Dallas, Texas, October 4–6, 1996, http://www.u.arizona.edu/ hameroff/water2.html

Hameroff, S. (1997a). "Funda-mental geometry: The Penrose–Hameroff Orch OR model of consciousness" in *Geometry and the foundations of science: Contributions from an Oxford conference honoring Roger Penrose*. Oxford Press (in press)

Hameroff, S. (1997b). "'More neural than thou': Reply to Churchland's 'Brainshy'" in *Toward a Science of Consciousness II – The 1996 Tucson Discussions and Debates*, ed. by S. Hameroff, A. Kaszniak & A. Scott, MIT Press, fall 1997

Hameroff, S. (1997c). "Did consciousness cause the Cambrian evolutionary explosion?" in *Toward a Science of Consciousness II – The 1996 Tucson Discussions and Debates*, ed. by S. Hameroff, A. Kaszniak & A. Scott, MIT Press, fall 1997

Hameroff, S. R. & Penrose, R. (1996a). "Orchestrated reduction of quantum coherence in brain microtubules: A model for consciousness" in *Toward a Science of Consciousness – The First Tucson Discussions and Debates*, ed. by S. R. Hameroff, A. Kaszniak & A. C. Scott, MIT Press, Cambridge, MA. Also published in *Mathematics and Computers in Simulation* 40: 453–480, http://www.u.arizona.edu/ hameroff/penrose1

Hameroff, S. R. & Penrose, R. (1996b). "Conscious events as orchestrated spacetime selections", *Journal of Consciousness Studies* 3(1): 36–53, http://www.u.arizona.edu/ hameroff/penrose2

Hameroff, S. R., Dayhoff, J. E., Lahoz-Beltra, R., Samsonovich, A. & Rasmussen, S. (1992). "Conformational automata in the cytoskeleton: models for molecular computation", *IEEE Computer* (Special Issue on Molecular Computing), 25(11): 30–39

Hameroff, S. R. & Watt, R. C. (1982). "Information processing in microtubules", *J. Theor. Biol.* 98: 549–561

Hameroff, S. R. & Watt, R. C. (1983). "Do anesthetics act by altering electron mobility?", *Anesth. Analg.* 62: 936–940

Hameroff, S. R., Rasmussen, S. & Mansson, B. (1989). "Molecular automata in microtubules: basic computational logic for the living state?" in *Artificial Life, the Santa Fe Institute Studies in the Sciences of Complexity, Vol. VI*, ed. by C. Langton, Addison-Wesley, Reading, MA, pp. 521–553

Herbert, N. (1993). *Elemental Mind: Human Consciousness and the New Physics*, Dutton/Penguin, New York

Hirokawa, N. (1991). "Molecular architecture and dynamics of the neuronal cytoskeleton" in *The Neuronal Cytoskeleton* ed. by R. D. Burgoyne, Wiley Liss, New York, pp. 5–74

Hotani, H., Lahoz-Beltra, R., Combs, B., Hameroff, S. & Rasmussen, S. (1992). "Microtubule dynamics, liposomes and artificial cells: in vitro observation and

cellular automata simulation of microtubule assembly/disassembly and membrane morphogenesis", *Nanobiology* 1: 61–74
Hunt, C. & Stebbings, H. (1994). "Role of MAPs and motors in the bundling and shimmering of native microtubules from insect ovarioles", *Cell Motility and the Cytoskeleton* 27: 69–78
Jibu, M. (1990). "On a heuristic model of the coherent mechanism of the global reaction process of a group of cells". *Bussei Kenkyuu* (Material Physics Research) 53(4): 431–436 (in Japanese)
Jibu, M. & Yasue, K. (1992a). "A physical picture of Umezawa's quantum brain dynamics" in *Cybernetics and Systems Research '92* ed. by R. Trappl, World Scientific, Singapore
Jibu, M., Hagan, S., Hameroff, S. R., Pribram, K. H. & Yasue, K. (1994) "Quantum optical coherence in cytoskeletal microtubules: Implications for brain function", *BioSystems* 32: 195–209
Kandel, E. R., Siegelbaum, S. A. & Schwartz, J. H. (1991). "Synaptic transmission" in *Principles of Neural Science* (Third Edition) ed. by E. R. Kandel, J. H. Schwartz, T. M. Jessell, Elsevier, NY, pp. 121–134
Karplus, M. & McCammon, J. A. (1983). "Protein ion channels, gates, receptors" in *Dynamics of Proteins: Elements and Function, Ann. Rev. Biochem.*, ed. by J. King, Benjamin/Cummings, Menlo Park, pp. 263–300
Kerwin, R. W. (1993). "Glutamate receptors, microtubule associated proteins and developmental anomaly in schizophrenia: A hypothesis" (editorial), *Psychological Medicine* 23: 547–551
Kim, H., Jensen, C. G. & Rebhund, L. I. (1986). "The binding of MAP2 and tau on brain microtubules in vitro" in Dynamic Aspects of Microtubule Biology, ed. by D. Soifer, *Ann. NY Acad. Sci.* 466: 218–239
Kirschner, M. & Mitchison, T. (1986). "Beyond self assembly: from microtubules to morphogenesis", *Cell* 45: 329–342
Kudo, T., Tada, K., Takeda, M. & Nishimura, T. (1990). "Learning impairment and microtubule-associated protein 2 (MAP2) decrease in gerbils under chronic cerebral hypoperfusion", *Stroke* 21: 1205–1209
Kwak, S. & Matus, A. (1988). "Denervation induces long-lasting changes in the distribution of microtubule proteins in hippocampal neurons", *J. Neurocytol.* 17: 189–195
Kwei, S., Jiang, C. & Haddad, G. G. (1993). "Acute anoxia-induced alterations in MAP2 immunoreactivity and neuronal morphology in rat hippocampus", *Brain Research* 620: 203–210
Lasek, R. J. (1981). "The dynamic ordering of neuronal cytoskeletons", *Neurosci. Res. Prog. Bull.* 19: 7–31
Lechleiter, J., Girard, S., Peralta, E. & Clapham, D. (1991). "Spiral calcium wave preparation and annihilation in Xenopus laevis oocytes", *Science* 252: 123–126
Lee, J. C., Field, D. J., George, H. J. & Head, J. (1986). "Biochemical and chemical properties of tubulin subspecies", *Ann. NY Acad. Sci.* 466: 111–128

Leggett, A. (1989). "Low temperature physics, supercondicutivity and superfluidity" in *The New Physics* ed. by P. Davies, Cambridge University Press, Cambridge, pp. 268–288

Leibniz, G. W. (1768). *Opera Omnia.* 6 volumes, ed. by Louis Dutens, Geneva

Lew, G. M. (1992). "Changes in microtubule-associated tau protein in human neuroblastoma cells after phencyclidine", *Gen. Pharmacol.* 23(3): 431–434

Lewis, S. A., Ivanov, I. E., Lee, G. H. & Cowan, N. J. (1989). "Organization of microtubules in dendrites and axons is determined by a short hydrophobic zipper in microtubule-associated proteins MAP2 and tau", *Nature* 342: 498–505

Libet, B., Wright, E. W. Jr., Feinstein, B. & Pearl, D. K. (1979). "Subjective referral of the timing for a conscious sensory experience", *Brain* 102: 193–224

Lloyd, S. (1993). "A potentially realizable quantum computer", *Science* 261: 1569–1571

Lynch, G. & Baudry, M. (1987). "Brain spectrin, calpain and long-term changes in synaptic efficacy", *Brain Res. Bull.* 18: 809–815

Marshall, I. N. (1989). "Consciousness and Bose-Einstein condensates", *New Ideas in Psychology* 7: 77–83

Mascarenhas, S. (1974). "The electret effect in bone and biopolymers and the bound water problem", *Ann. NY Acad. Sci.* 238: 36–52

Matsumoto, G. & Sakai, H. (1979). "Microtubules inside the plasma membrane of squid giant axons and their possible physiological function", *J. Membr. Biol.* 50: 1–14

Matsuyama, S. S. & Jarvik, L. F. (1989). "Hypothesis: Microtubules, a key to Alzheimer's disease", *Proc. Nat. Acad. Sci.* 86: 8152–8156

Melki, R., Carlier, M. F., Pantaloni, D. & Timasheff, S. N. (1989). "Cold depolymerization of microtubules to double rings: geometric stabilization of assemblies", *Biochemistry* 28: 9143–9152

Mileusnic, R., Rose, S. P. & Tillson, P. (1980). "Passive avoidance learning results in region specific changes in concentration of, and incorporation into, colchicine binding proteins in the chick forebrain", *Neur. Chem.* 34: 1007–1015

Moshkov, D. A., Savaljeva, L. N., Yanjushina, G. V. & Funtikov, V. A. (1992). "Structural and chemical changes in the cytoskeleton of the goldfish Mauthner cells after vestibular stimulation", *Acta Histochemica Suppl-Band* XLI, S: 241–247

Neubauer, C., Phelan, A. M., Keus, H. & Lange, D. G. (1990). "Microwave irradiation of rats at 2.45 GHz activates pinocytotic-like uptake of tracer by capillary endothelial cells of cerebral cortex", *Bioelectromagnetics* 11: 261–268

Penrose, R. (1971). in *Quantum Theory and Beyond* ed. by E. A. Bastin, Cambridge University Press, Cambridge, U.K.

Penrose, R. (1989). *The Emperor's New Mind*, Oxford University Press, Oxford

Penrose, R. (1994). *Shadows of the Mind*, Oxford University Press, Oxford

Penrose, R. (1996). "On gravity's role in quantum state reduction", *General relativity and gravitation* 28(5): 581–600

Penrose, R. (1997). *The Large, the Small and the Human Mind*, Cambridge University Press, Cambridge, U.K.

Penrose, R. & Hameroff, S. R. (1995). "What gaps? Reply to Grush and Churchland", *Journal of Consciousness Studies* 2(2): 99–112

Popp, F. A. (1986). "On the coherence of ultraweak photoemission from living tissue", in *Disequilibrium and self-organization* ed. by C. W. Kilmister, D. Reidel Publishing, Dordrecht and Boston

Puck, T. T. (1987). "Genetic regulation of growth control: Role of cyclic AMP and cell cytoskeleton", *Somat. Cell Mol. Genet.* 13: 451–457

Rasmussen, S., Karampurwala, H., Vaidyanath, R., Jensen, K. S. & Hameroff, S. (1990). "Computational connectionism within neurons: A model of cytoskeletal automata subserving neural networks", *Physica D* 42: 428–449

Reinsch, S. S., Mitchison, T. J. & Kirschner, M. (1991). "Microtubule polymer assembly and transport during axonal elongation", *J. Cell Biol.* 115(2): 365–379

Roth, L. E. & Pihlaja, D. J. (1977). "Gradionation: Hypothesis for positioning and patterning", *J. Protozoology* 24(1): 2–9

Rovelli, C. & Smolin, L. (1995a). "Discreteness of area and volume in quantum gravity", *Nuclear Physics* B 442: 593–619

Rovelli, C. & Smolin, L. (1995b). "Spin networks in quantum gravity", *Physical Review* D 52(10): 5743–5759

Sabry, J. H., O'Connor, T. P., Evans, L., Toroian-Raymond, A., Kirschner, M. & Bentley, D. (1991). "Microtubule behavior during guidance of pioneer neuron growth cones in situ", *J. Cell Biol.* 115(2): 381–395

Samsonovich, A., Scott, A. & Hameroff, S. R. (1992). "Acousto-conformational transitions in cytoskeletal microtubules: implications for neuro-like protein array devices", *Nanobiology* 457–468

Shimony, A. (1993). *Search for a Naturalistic World View – Volume II. Natural Science and Metaphysics*, Cambridge University Press, Cambridge, U.K.

Silva, A. V., Stevens, C. F., Tonegawa, S. & Wang, N. (1992). "Deficient hippocampal long-term potentiation in a-calcium-calmodulin kinase II mutant mice", *Science* 257: 201–206

Spinoza, B. (1677). *Ethica in Opera quotque reperta sunt*, 3rd edition, ed. by. J. van Vloten & J. P. N. Land, Den Haag

Surridge, C. D. & Burns, R. G. (1992). "Phosphatidylinositol inhibits microtubule assembly by binding to microtubule-associated protein 2 at a single, specific, high affinity site", *Biochemistry* 31: 6140–6144

Tabony, J. & Job, D. (1992). "Microtubular dissipative structures in biological auto-organization and pattern formation", *Nanobiology* 1(2): 131–147

Taft, W. C., Yang, K., Dixon, C. E., Clifton, G. L. & Hayes, R. L. (1993). "Hypothermia attenuates the loss of hippocampal microtubule-associated protein 2 (MAP2) following traumatic brain injury", *J. Cerebral Blood Flow and Metabolism* 13: 796–802

Tan, S. E., Abel, E. L. & Berman, R. F. (1993). "Brain MAP2 phosphorylation is decreased following prenatal alcohol exposure in rats", *Alcohol* 10: 391–396

Theurkauf, W. E. & Vallee, R. B. (1983). "Extensive cAMP-dependent and cAMP-independent phosphorylation of microtubule associated protein 2", *J. Biol. Chem.* 258: 7883–7886

Tsukidate, K., Yamamoto, K., Snyder, J. W. & Farber, J. L. (1993). "Microtubule antagonists activate programmed cell death (apoptosis) in cultured rat hepatocytes", *American J. Pathol.* 143(3): 918–925

Tuszynski, J., Hameroff, S., Sataric M. V., Trpisova, B & Nip, M. L. A. (1995). "Ferroelectric behavior in microtubule dipole lattices; implications for information processing, signaling and assembly/disassembly", *J. Theor. Biol.* 174: 371–380

Varela, F. J., Thompson, E. & Rosch, E. (1992). *The embodied Mind – Cognitive Science and Human Experience*, MIT Press, Cambridge, MA

Vassilev, P., Kanazirska, M. & Tien, H. T. (1985). "Intermembrane linkage mediated by tubulin", *Biochem. Biophys. Res. Comm.* 126: 559–565

Vernon, G. G. & Woolley, D. M. (1995). "The propagation of a zone of activation along groups of flagellar doublet microtubules", *Experimental Cell Research* 220(2): 482–494

Von Neumann, J. (1966). in *Theory of self-reproducing automata*, ed. by A. W. Burks, University of Illinois Press, Urbana

Vos, M. H., Rappaport, J., Lambry, J. Ch., Breton, J. & Martin, J.L. (1992). "Visualization of coherent nuclear motion in a membrane protein by femtosecond laser spectroscopy", *Nature* 363: 320–325

Wang, N. & Rasenick, M. M. (1991). "Tubulin-G protein interactions involve microtubule polymerization domains", *Biochemistry* 30: 10957–10965

Watterson, J. G. (1987). "A role for water in cell structure", *Biochem. J.* 248: 615–617

Wheeler, J. A. (1990). "Information, physics, quantum: The search for links" in *Complexity, Entropy, and the Physics of Information*, ed. by W. Zurek, Addison-Wesley

Whitehead, A. N. (1929). *Process and Reality*, Macmillan, NY

Wolfram, S. (1984). "Universality and complexity in cellular automata", *Physica D* 10: 1–35

Wuensche, A. (1992). "Basins of attraction in disordered networks" in *Artificial Neural Networks*, 2, ed. by I. Aleksander & J. Taylor, Elsevier, Amsterdam, pp. 1325–1344

Wulf, R. J. Featherstone, R. M. (1957). "A correlation of Van der Waals constants with anesthetic potency", *Anesthesiology* 18: 97–105

Yagi, T., Kamimura, S. & Kamiya, R. (1994). "Nanometer scale vibrations in mutant axonemes of Chlamydomonas", *Cell Motility and the Cytoskeleton* 29: 177–185

Yu, W. & Baas, P. W. (1994). "Changes in microtubule number and length during axon differentiation", *J. Neurosci.* 14: 2818–2829

6. Of Maps and Territories: A Three Point Landing on the Mind-Body Problem

Michael Conrad and Deborah Conrad

6.1 Introduction

The relationship between mentality and the material world of physics and chemistry is arguably the pivotal question that spans philosophy and the natural sciences. For some the issue is a nonissue: mentality is an illusion; for others it is also a nonissue: the material world is an illusion. One currently popular view, the so-called computational cognitive paradigm, takes its cue from artificial intelligence (AI) and connectionist neural network theory. According to this school, mind is a manifestation of the signals flowing among simple switches. The material nature of the signals and the switches is no matter. According to the classical artificial intelligence paradigm, programs run on a silicon digital computer operated in a serial mode could possess mentality. According to the connectionist neural network approach mentality can inhere in a collection of silicon switches that are appropriately linked up or in a collection of virtual switches that are simulated on a serial machine. The carbon nature of the brain, according to these views, is incidental.

The thesis to be argued in this chapter is quite the opposite. The physical-chemical-dynamical nature of organisms and the universe is inextricably tied to mentality. Our argument is three legged:

1. *Analysis of rules* (D. Conrad 1988, 1993). The cognitive activities of organisms, in particular rule-following behavior, raise awkward problems for the computational cognitive paradigm. Choice and intention enter in a fundamental way.
2. *Tradeoff theorem* (M. Conrad 1985, 1988). Analysis of programmable digital computers and connectionist models shows that they do not have the real-time computational capabilities exhibited by organisms. Nonprogrammable, evolvable systems are necessary for such capabilities, and only carbon-based systems fill the bill.

3. *Great chain of being brain models* (M. Conrad 1987, 1989, 1993). Nonprogrammable systems of the type required, because of their openness to evolution, are physics dependent. Fundamental features of the physical world, in particular wave function collapse, are relevant to their capabilities. The type of choice and intention required for rule following behavior thus has a place in a materials-dependent view that it does not have in the purely algorithmic view.

The formal maps required to describe behavior that can be characterized by everyday rules, so the argument will go, are so complicated that they cannot exist in any operational sense apart from the material nature of the systems that exhibit this behavior. Think of a map that is so embedded in the territory it describes that it cannot be separated from it. It would lose in some measure the quality of "mapness", but at the same time would gain some properties of "territoriness" that could never be discerned in a pure map. We don't claim that our argument is knockdown. The mind/brain problem may never admit an analysis that forces the whole world to a justified consensus. We do claim that the three legs of the argument fit together to form a stable stool on which to stand, that the AI and connectionist paradigms based on current computer technology fail this requirement, and that the embedded map metaphor may therefore have some heuristic value.

6.2 Everyday Rules and Behavior

"Mentality" and "mind" cover much diverse and controversial territory, charted but by no means thoroughly explored. A classification of mental kinds is beyond the scope of this chapter, but two broad types of mentality should be distinguished: cognition and consciousness. Cognition has to do (roughly) with learning, reasoning, knowing, and remembering; consciousness with being occurrently aware of one's environment, one's own body or one's own mental state. Both are mental, if anything is, and in many kinds of intelligent human activity both are present; however, the mental capacities that will be highlighted in this discussion of rule following are primarily cognitive ones. An oft made criticism of AI and connectionist claims to explain mentality is that their models do not account for consciousness (Haugeland 1981; Searle 1984). Our argument in this chapter is directed both at this shortcoming and at the apparently better supported claims on behalf of a materials-independent explanation of cognition.

Rule following represents only a part of a complex group of human practices that includes rule formulating, rule interpretation, rule bending and rule breaking, rule adoption and rule change over time. The term "rule following" may suggest rigidity and predictability of behavior, but strict compliance with precisely specified directives is only one relation that human behavior can have to rules. This brief discussion concentrates on aspects of "everyday" rule practices, particularly rule following, that involve choice and intention.

To follow a rule is to do what it tells you to do ("Check bags at the front of the store") or refrain from doing what it forbids you to do ("Keep off the grass"). The behavior which is enjoined, permitted, or forbidden by a rule is the type of behavior it "governs", or applies to; analogously a rule is also said to govern the group of people whose behavior it is intended to regulate. Thus the above rules regulate grass-treading behavior and bag-carrying behavior, respectively, in areas where they are posted. The exact behavior and group a rule governs at any given time are determined by background and contextual elements, e. g., by the shared intentions of the community in which the rule is operative, and are often not made explicit in the rule. For instance, it is generally understood that the rule in stores about checking bags does not apply to women's purses and that gardeners are allowed to step on the grass, though these exceptions are not part of the rule formulations themselves.

Rules could be formulated more precisely and completely, but there are distinct advantages to leaving some parameters open to interpretation. For one thing, it allows individuals in particular rule governed situations flexibility to choose in precisely what manner to comply with a rule, in light of current community expectations concerning the rule-governed behavior. For example a well known traffic regulation says "Maximum speed for motor vehicles on divided highways is 55 miles per hour; minimum speed is 45 mph." A literal reading of the rule might lead to the conclusion that a person is obliged to drive between 45 and 55 mph under all circumstances, since no narrower set of circumstances is specified. But well informed drivers know that the more general rule to drive at a speed suitable for road conditions, including the speeds of other vehicles, takes precedence over this rule and sometimes dictates that one drive even slower than 45 mph. Under very few circumstances would the rule as generally understood allow driving above the maximum speed limit. But given our shared knowledge of the rationale and intended effect of the speed limit, there will tend to be agreement about the sorts of cases that would constitute valid exceptions.

Such openness of formulation also accommodates gradual change in circumstances over time without the necessity of repeated reformulation of rules. The regulation above does not list all the types of motor vehicles which fall under it, thus allowing the speed limit to apply without revision of the rule to new types of motor vehicles as they are designed and appear on the road. Community intentions as to what behavior counts as compliance with the rule can also evolve by degrees within the latitude afforded by incomplete formulation.

The parameters of a rule that are almost always specified are a type of behavior and whether it is forbidden, required, or permitted. But in the limiting case there can even be rules operative in a group that are not explicitly formulated at all. The existence of such rules is indicated by patterns of approval and disapproval of particular forms of behavior, along with (informal) sanctions for noncompliance. For instance in most human societies there is an unwritten rule that people not infringe upon one another's "personal space." Thus, when talking to someone, an adult is expected to stand a certain minimum distance away from that person. The distance is not specified in any explicit rule, but a visitor to a culture other than his own that approves of a different distance becomes aware that a different rule is operative. (Either someone stands too close to him, and he automatically moves back, or vice-versa.) When they are too close to someone else, young children are admonished to move further away. Generally, no rule is quoted, but the pattern of reminders eventually produces behavior in accordance with the implicit rule.

Such everyday rules, as opposed to rules that are communicated to computers, do not determine the behavior they govern; that is, there is not invariable compliance with a rule, even when a person is in a situation in which the rule applies, understands what behavior it is that the rule calls for, and is capable of suitably adjusting his behavior to accord with the rule. Think of civil disobedience, laziness, or weakness of will, to name only a few of the intentional causes of failure to act in accordance with rules. The possibility of choosing what to do in light of what a rule prescribes is clearly presupposed by "permissive" rules – those that neither enjoin nor prohibit a type of behavior, but rather grant permission to engage in it. Thus in chess a player may castle the king once during a game if certain conditions are met and he decides to do so; in some buildings there are areas where people may smoke if they choose to; the WALK signal, though it reads like a command, grants permission to cross the street. In such cases the choices are licensed by rules that grant the requisite permission.

Besides cases of intentional rule breaking unintentional failure to comply with everyday rules provides evidence that rules "govern" behavior in a very complex way. People can misunderstand or misinterpret rules; or they may try to follow a rule but lack the mental or physical skills to succeed. In speaking a language failure to accord with some of its rules is often due to haste or inattention. This has been noted by linguists who record the utterances of mothers in the presence of young children in the course of language acquisition studies. The fact that so many of the utterances are not "according to Hoyle" (i. e., not in accord with the rules of grammar that children eventually acquire) led Chomsky and others to argue that some of the general rules of language must be innate. For the same reason in the construction of (computer) speech recognition systems attention must be paid to the circumstance that "speech is remarkable for the variety of rules it follows and even more remarkable for the rules it violates" (White 1990). Sometimes these linguistic rules are explicit ones, found in grammar books and taught in school. Others are implicit, being exhibited in the behavior of most members of a community most of the time in the situations that these implicit rules govern.

A final example of the pivotal role of choice and intention is the use of the so-called rules of conversational implicature (Grice 1975; Levinson 1983) to facilitate communication. The most general formulation of these (predominantly implicit) rules is that a conversant's contribution to a conversation should be appropriate, given what has gone before and the accepted purpose or direction of the conversation. More specifically, a person's utterance should be true as far as he knows; as informative as necessary in the circumstances, but no more informative; relevant to the purpose or direction of the conversation; and clear.

The interest in conversational implicature is not that speakers always comply in a straightforward way with every one of these rules. Sometimes reaching the overall goal of communication is facilitated by the apparent violation of one or more of them. For example, if I ask you, "Can you tell me the time?" and you reply, "Well, the mail carrier has come, " it appears that you have violated the rule that your contribution to any conversation should be relevant. But if I assume that you intend (are trying) to co-operate in furthering the goals of the conversation, I will be prompted to try to figure out how what you did say could be construed as relevant to answering my question. I can use my knowledge of the rules of conversation (say as much of what you know that is relevant) and the assumption that speakers are usually attempting to follow them to infer that you don't know the exact time, but you do know something that is relevant to the time of day, namely that it is past the time when the mail carrier usually comes. In other words, I use the apparent breaking of a rule (what you said isn't a direct answer to my question) as a signal to look for the way in which you intend it to be relevant.

The examples discussed here do not exhibit particularly obscure or controversial aspects of everyday rules, but they contrast sharply with the nature of rules that play a role in AI and connectionist theory. Such computational rules produce behavior in digital computer models and (simulated or hardwired) neural nets that is rule governed in a fundamentally different way. Choice in the sense of deciding whether or not to execute a rule when all other conditions for doing so are met is not a concept that applies to digital computer programs or to patterns of connections between nodes in a neural net. If the logical conditions (which are ultimately physical conditions) for executing an instruction are fulfilled when a given instruction is encountered, the instruction is executed (barring technical problems with the machine running the program); if the conditions are not fulfilled, it branches to another program instruction, and continues to operate in this fashion. Rules communicated to computers are explicit and fully specified; there is no need for context dependent interpretation or possibility of flexible compliance guided by recognition of the designer's (or anyone's) intentions. Of course these are highly inappropriate categories to apply to the lowest level of computer rules, given the design constraints that programmable systems must satisfy. But at some level, presumably that of simulation, capacities involved in acting on everyday rules must be exhibited by computers or neural networks in order to validate the AI and connectionist view of the relation between mind and matter. The constraints on programmable and connectionist systems as well as the prospects for successful simulation of everyday rule following behavior will be discussed in the next section.

Why the emphasis on rule following in a debate about the nature of the mental? It is, to be sure, only one of myriad human intellectual capacities and only one of those that have essentially semantic and interpretative components. Any symbolism, whether it is used to express poems, physical laws, computer programs or everyday rules, requires interpretation. No symbol means anything intrinsically, i. e., apart from the conventional assignment of its use or meaning. This is the thrust of Wittgenstein's well known "arrow" example (Wittgenstein 1953). Though an arrow, say printed on a sign, seems to require no interpretation, it is solely convention that makes it point in the direction of its head rather than in any other, e. g., in the direction of its tail. This is in contrast to a material arrow, e. g., one shot from a bow, that requires no interpretation in order to point (or fly) in a given direction.

Everyday rules, due to their wide variety and routine incompleteness of formulation, require a second level of highly context dependent interpretation, where context involves an understanding of a web of cultural practices as well as details of the specific situation in which they are being applied. In contrast it is (a restricted kind of) rules as woven into computer programs and exhibited by the connections in neural nets that bear the burden of explanation of mental capacities in the main alternative to the materials-specific view defended here. So the nature and potential of rules is appropriately at the cynosure of the debate.

6.3 The Trade-off Principle: Antithetical Ways of Tying Rules to Matter

Now let us consider whether rule following behavior of the type described above could be exhibited by programmable machines. If this were possible it would mean that mind is essentially independent of physics and the material sciences, for the essence of programmable (digital) computers is that they are physical realizations of formal systems. Today all such systems are constructed from inorganic materials, in particular silicon. But this is because these materials are particularly well suited for constructing reliable switches that can be wired together in such a way that the user can embed desired rules. Furthermore it is possible for the user to do this without performing any experiments or calculations that draw on physics or chemistry. All that is necessary is that he use a simple user manual that specifies precisely the input-output behavior of the switching components. We will call systems of this type structurally programmable, since the rules that generate their behavior are encoded in the states of the components and the connections among them according to a finite user manual (Conrad 1983, 1988).

First it should be evident that rules of the above type are more limited than the everyday rules described in the previous section. There is a wide variety of formalisms for expressing computational rules: Turing machines, recursive functions, Post's productions, or any of a large number of high level languages (Pascal, Lisp, ...). All of these different string processing formalisms are equivalent in terms of the class of functions they compute (in the absence of restrictions on available space or time).

Because of this equivalence we can describe the type of rules that generate the behavior of digital computers in terms of any one of these formalisms. Let us choose Turing machines, since these are particularly perspicuous so far as the putative connection to mind is concerned. Thus we can imagine (as Turing seems to have done) a child doing arithmetic on a note pad (Turing 1937). The child has a finite set of discrete "states", putatively identified with states of mind. He can read and write symbols on a pad that is divided up into squares, using a finite alphabet of discrete symbols, and can move his "read-write" head from square to another. It is sufficient to model the pad as a one-dimensional tape and to suppose that the moves of the read-write head are restricted to one of three possibilities: one tape square to the right, one tape square to the left, and no move. The program of a Turing machine can then be written as a set of rules of the form

$$q_i x_j y_k m q_f$$

where q_i is the state of our discretized child at the beginning of the computational process, x_j is the input symbol on the tape square at which he is looking, y_k is the symbol he writes on this square, m is the move he makes, and q_f is his state of mind at the end of the process. As indicated above a set of such elemental process descriptions, or Turing quintuples, defines the program of the Turing machine. The program is a rule built up from the elementary rules. It can be regarded as a (possibly) giant pair of tables, one for the output given the present state and input and one for the next state given the present state and input. Each of the tables is a function, or rule, in the precise sense that it maps arguments into values. (The initial arguments are the initial configurations on the input tape. The value is the configuration when the machine comes to a halt, if it does come to a halt; if it does not the function is undefined).

The term "Turing machine" is actually an abuse of language, since what we are dealing with is a rule and not an actual machine. This rule requires a material implementation to actually generate, describe, or exhibit a computational process. The implementer might be a human being who reads and follows the rule. Or it might be an actual physical device that realizes the finite control, the read-write head, and tape. This distinction is extremely important for our purposes, for it is quite different to say that the everyday behavior of a human being is generated by a rule and to say that we can implement this rule in materials such as silicon that are of a different nature than those from which biological organisms are built.

The Turing quintuples outlined above are clearly narrower in scope than everyday rules. It is possible to state many everyday rules with a short sequence of symbols that could not be reexpressed in terms of a single Turing quintuple or in terms of a short Turing program. But this does not mean that such rules could not be expressed in terms of elaborate Turing programs. After all, rules stated in a high level programming language, such as Pascal, in general require somewhat elaborate Turing programs to duplicate (or simulate). The most challenging case is perhaps the permissive rule – you can cross on the WALK signal. Conceivably it would be possible to use Turing rules, or any equivalent computational formalism, to create an artificial world with artificial human organisms. The rules governing the organisms might lead them to cross on the WALK, or not to cross, only if a variety of conditions are met. Whether they cross could in addition be controlled by a random (in practice pseudo-random) number generator that confers some stochasticity. Whether the model organisms have intentions or mental experiences would not be known. But since we don't observe the mental states of other human beings, only their behavior, the situation on the surface appears not fundamentally different for computer organisms and fellow humans. The argument then goes, at least for the artificial intelligence and artificial life enthusiasts, that the model organisms have as much right to be regarded as cognitive, even conscious, entities as humans have.

This issue has been argued and counterargued since the inception of digital computers. It is connected to the claim of the universality of digital simulatability, sometimes known as the strong version of the Turing–Church thesis. The claim is that everything that is effectively computable is computable by Turing machines. The weak interpretation of this claim is that no string processing formalism more powerful than Turing machines will ever be discovered. The strong interpretation is that all processes in nature are digitally simulatable, or duplicatable up to a reasonably simple coding, apart from practical limits connected with space and time bounds. For our purposes it is the strong interpretation that is of interest, since it is the one that bears on the mind-matter relationship. Thus suppose that in nature some system existed that could compute some function not in the class of Turing computable functions. This system could be adjoined to a digital computer to effectively solve problems that could not be solved by Turing machines, even though it might not do so in a string-like way. The strong version of the Turing–Church thesis asserts that the processes of mind do not rely on any such extra-Turing processes.

Let us suppose that the strong version of the Turing–Church thesis is true. Would this mean that it is in principle possible to create an artificial world of the Turing type in which model organisms could exhibit intelligence comparable to human intelligence, as ascertained by their ability to act as if they were following everyday rules? Turing raised a question of this sort – though in the context of a much less demanding task – in his famous Turing test. He supposed that it would be possible to program a digital computer in such a way that it would be indistinguishable from a flesh and blood human, assuming that the digital machine and the human were put behind a curtain that hides their material characteristics. Such a machine, Turing claimed, could be called intelligent with just as much justification as the human is called intelligent.

The problem with this claim is the tacit assumption that it is possible to actually build a machine – a material instantiation of a Turing program – that could pass the Turing test. All that Turing–Church implies is that systems in nature, including the systems responsible for human intelligence, are describable by mathematical maps that are Turing computable. There is no claim that the amount of time and space required to compute the map satisfies real-time constraints imposed by the environment, or that they are even computable given all the time, space, and energy available in the universe. This consideration did not weigh very heavily early in the rather short history of computer science, since digital machines execute simple string processing operations much more rapidly than humans. But when it comes to recreating an artificial world that can duplicate the kind of rules that humans easily and rapidly handle, and to do so in a comparably vast variety of contexts, the question as to whether it is physically possible for the maps to be executed by a digital computer and indeed even be written down and communicated to it becomes important.

At this point we can draw on the trade-off principle, but without going into formal or quantitative details (see Conrad 1990). The basic statement is that no information processing system can be programmable at the structural level (like a digital machine) and at the same time have high computational efficiency and evolutionary adaptability. Recall that structural programmability captures the idea of a finite user manual. A system is structurally programmable if the program that generates its behavior is encoded in the states of its components and in their connections according to a program which is discrete, finite in nature, and precise (no approximation or elements of continuity). For this to be the case the input-output behavior of each component of the system must also conform to simple user manual rules. When new components are added constraints must be present that preclude new interactions from occurring. The conflict between structural programmability and efficiency is due to the fact that the vast majority of interactions that potentially occur among the particles of the system must be frozen out in order to achieve this condition. As the system gets bigger in terms of number of components the fraction of interactions actually used decreases rapidly. This is true even if the system is run in a parallel mode. The situation is worse for serial machines – the only ones that can in general be effectively programmed without explosive compilation costs – since only one or a few of the components are active at any given time.

The conflict between structural programmability and evolutionary adaptability follows from the fact that the behavior generated by a program is generally very sensitive to slight changes in the program, apart from changes in parameters. Any system that has a program encoded in its structure will inherit this structure-function fragility. According to Turing-Church it should be possible to use structurally programmable machines to create virtual (or simulated) structure-function plasticity, for example by writing programs to represent the nonprogrammable structure-function relations of organisms. This in fact can be done but at enormous computational cost.

It is quite possible to build up artificial world models – brain models and evolution models, for example (Conrad 1989; Rizki & Conrad 1989). But for the reasons indicated above the amount of processing that the digital computer is capable of is inordinately smaller than that of the natural systems modeled. Even a small molecule treated classically generally swamps the resources available in supercomputers unless simplifying assumptions are introduced that greatly limit the range of conditions over which the computation is applicable. Biological systems, because they ultimately learn through the opportunity seeking process of variation and natural selection, are free to exploit such computationally refractory interactions for coherent functions of the type described by everyday rules, whereas digital computers are excluded from doing so.

This conclusion extends with only minor modifications to connectionist architectures. The trade-off principle still applies, since connectionism as usually conceived postulates that intelligence is in the pattern of connections, not in the components. Alternatively stated, a connectionist system is a structurally programmable system run in a parallel mode. Artificial neural nets, for example, consist of neurons each of which has a simple, precise user manual definition. Consequently the vast majority of the potential interactions among the components must still be suppressed by the engineer. The only significant difference from the serial digital computer is that many of the components can be active at once, i. e., high parallelism is in principle possible. The Turing quintuples can still be used to represent the behavior of the network (or approximate it if it is a continuous system). The number of components required to implement the system is for suitable problem domains decreased, since each state is represented by the activity of a combination of components. The inputs that the network can "read" in one time interval can be more complicated as well.

For problem domains that map into pattern recognition and effector control the potential efficiency of connectionist neural networks is increased over serial machines, but is less than in systems that do not rigidly restrict the potential interactions. The capacity of the system to learn through changes in the structure of the network is also increased; indeed this is essential because the effective use of parallelism typically requires the dynamics of the system to be amenable to molding through learning algorithms. This is what limits the domain applicability. To achieve the requisite learnability it is generally necessary to ensure that variations in the structure will leave the system well formed (or to impose differentiability conditions that in effect serve the same purpose). But the class of functions computable by a system so restricted is very much smaller than the Turing computable functions. In practice it is difficult to represent high level knowledge or to deal with symbolic information processing. Everyday rules are for the most part out of the problem domain.

What the trade-off principle says is that the maps that organisms realize are tied to them in a different way than are the maps that digital computers realize, and that this is so even under the somewhat dubious assumption that such maps are Turing computable. Digital computer maps are explicitly written into their structure, at great cost in terms of computational efficiency and evolutionary adaptability. This is also true for maps that describe the behavior of connectionist architectures, with the difference that the embedding algorithms are much more complicated. In both cases the maps can be ported from one machine to another in a materials-independent way, provided that the machine supports switching functions that make it structurally programmable and computation universal. This has led some to conclude that for every kind of process the map is the crucial thing and the material particularities of the realization are irrelevant. But this embedded encoding relation is unique to digital computers. The falling apple does not encode Newton's equation in its structure – if it did, oranges with their different structure would fall according to a different gravitational law than apples. This is why a falling apple is so much more efficient at behaving like an apple than a digital computer is at simulating it. The rule governing the fall is nevertheless simple enough to be written down and communicated to a digital computer. But this is not always feasible and sometimes not possible even in principle. A universe is so much more efficient at behaving the way it does than any digital computer could conceivably be that no other system could realize the map that describes its time development. The map in this extreme case becomes inseparable from the territory. The maps that generate (or describe) the behavior of organisms are tied to them in a similarly inseparable manner, at least so far as satisfying real time requirements is concerned. This follows from the fact that they are the products of evolution, *a fortiori* they must have high evolutionary adaptability. According to the trade-off principle this precludes structural programmability. But it is compatible with high computational efficiency. The maps, we shall argue shortly, are so computationally demanding when measured in terms of Turing machine operations that they could never be implemented by digital machines, and indeed never be embodied in any adequate way in physical-chemical media other than carbon.

One might of course choose to think of the map as existing as an abstract entity, and it might even be possible to partially express it in some kind of general symbolism. Our point is that operationally it may not be possible to communicate it to the programmable environment of a digital computer and have the full behavior of the matter that it is supposed to describe duplicated on an acceptable time scale. The hypothetical map (or program) may be small enough to fit into the computer, but too resource consuming to generate an adequate approximation to the system simulated to establish its validity, or it may even be too big to describe on an input medium of acceptable size. It is in this operational sense that we say that the map of such a system is tied to the territory. The territory is a given. The existence of the map and its manner of existing is much less clear. But the difference in efficiency between even fairly simple systems and computer simulations of them indicates that in many cases – and possibly all cases of interest from the standpoint of cognitive behavior of the everyday sort – such a map is not in general portable from the matter whose behavior it putatively describes or generates, for porting it would require so much of the universe to be converted to programmable machinery that the original object system would have to be destroyed.

We can think, metaphorically, of a world which is made out of heavy and light material (lead and paper). We can draw a map of this world on paper without interfering with it very much. But if our world were made only out of lead the map of it would also have to be made out of lead. A complete map would inevitably alter the territory. The trade-off principle imposes much more severe constraints on mapmaking. The map must not only be embodied in materials that actually exist in the world, but be embodied in a form that in general requires it to use many more material resources than the object system in order to duplicate the behavior of this system. (As stated above, one could choose to view the map as having a Platonic reality above the shapes and features of the world. But to actually represent and manipulate it requires operations in this world, with the material of this world, and thus from the operational point of view the map does not exist as a truly separate entity. The true Platonist might evade this operational requirement by hypothesizing that it is possible to answer questions about systems in this world by actually contacting the abstract realm in which he supposes the map to reside, but clearly such a position would completely deviate from the Turing-Church thesis and from the assumption that it is possible to construct artificial intelligence.)

6.4 The Great Chain of Being Model

Now let us turn to the third leg of our argument – that biological organisms in fact process information in a manner that exploits physical-dynamical features, including quantum features, that are computationally costly when measured in terms of the number of digital switching operations required, and that in some cases may not even be digitally simulatable. These quantum features lead to a concept of state that is richer than the classical state of the Turing quintuples and that more naturally includes a place for mental qualities, such as the (conscious or unconscious) intentions that we associate with permissive rules, or indeed with any rule that we can choose not to follow.

To keep the line of argument in mind we recall that we started with a rule phenomenology that was far more general than the types of rules that occur in computer programs. We then argued that in many cases realizing such everyday rules with digital machines is blocked in practice by the trade-off principle, at least as far as satisfying real-time constraints is concerned, since computationally demanding artificial world programs are required to represent them.

Biological systems, because they are evolvable and nonprogrammable, are not blocked in this way. As noted above, it would be unreasonable to expect evolution to fail to exploit accessible physical-dynamical effects simply because this would complicate the work of the scientist who wants to write down a set of rules to explain the behavior of an organism. Or put more positively, it is reasonable to expect that whatever biological systems can usefully couple to they will draw into the chain of events that underlie their functional capabilities, regardless of the space or time scale of these events.

Let us consider neurons, since these are the cells most intimately connected with intelligence. The classic picture of the nerve impulse is that it is purely a membrane phenomenon, only indirectly supported by the energy metabolism of the cell, and hence for some reason isolated from microphysical effects. It turns out that this is not the case even for primary neuroreceptor cells, such as retinal cells. The impinging photon triggers the production of internal biochemical changes that control ion channels, and hence control the input-output behavior of these neurons. The nerve impulse activity of at least some central nervous system neurons is now also known to be controlled by internal chemical and molecular events. Second messengers triggered by incoming impulses act on target proteins (kinases) that in turn act on effector proteins, such as channel proteins in the membrane (Liberman et al. 1985). A wide variety of mechanisms, including cytoskeletal signal processing, may be involved (cf. Hameroff 1987; Conrad 1990)

The neuron in the classic picture could be modeled as some kind of threshold element whose behavior is for all practical purposes independent of its carbon macromolecular nature. Accordingly the brain could be approximated as a structurally programmable system. The discovery of second messenger systems that link the relatively macroscopic level of membrane electric activity to the mesoscopic and microscopic levels of molecular function renders this picture untenable. The neuron is better pictured as a transduction-amplification module that transforms spatiotemporal patterns of input to temporal patterns of output. The input patterns are transduced to microscopic representations, processed at the molecular level, and then amplified back to the more macroscopic level of nerve impulses. The brain may be conceived of as a network of special purpose pattern transformers linked together for the performance of coherent function. The system, since it is structurally nonprogrammable, is well suited for molding through phylogenetic evolution and ontogenetic learning processes.

We will not here consider the types of neural architectures and learning algorithms that are appropriate to this picture, or the manner in which it is possible to manipulate complex assemblies of nerve cell activity to produce associative structures, to orchestrate groups of neurons suitable for particular perception-action tasks, and to utilize self-representations (see Conrad et al. 1989). The more pertinent question for the present purposes is directed to the inside of the neuron, since it is here that the higher level control processes make their inseparable contact with the properties of matter.

The key point is specificity, the ability of macromolecules to bind with or act catalytically on other molecules in a manner that is highly selective and that can at the same time be fast. The term molecular recognition, or molecular pattern matching, reflects this high degree of specificity. Conformation-driven processing is another good term, since complementary shape fitting plays a prominent role. The catalytic proteins, or enzymes, provide the clearest and most important example. At first glance one might regard enzymes as switches amenable to a simple logical description. But the main feature of enzymes is that they exert their switching action only on specific bonds in specific object molecules. The classic metaphor is that of a lock and key. The key switches the lock from one state to another; but the more important fact about keys is that they work only on specific locks. The geometry, and in the case of enzymes the dynamic changes in geometry, converts what would be a difficult pattern recognition problem for a system of simple switches into a physical process whose achievement is as natural, even unavoidable, as the fall of an apple in a gravitational field.

Such conformational processing accompanies virtually all the biochemical processes that underlie the functionality of neurons and other cells. But to highlight the contribution of recognition to the pattern processing capability of the cell as a whole, and to isolate it from other important processes such as catalytic switching and diffusion, let us consider the gedanken device illustrated in Fig. 6.1. This is a model cell with receptors (input lines) and output. If a receptor is activated a specific macromolecule is released. The pattern of impinging signals is thus transduced to a pattern of specific molecules. These self-assemble to form a polymacromolecular mosaic whose shape depends uniquely on the input pattern. The cell produces an output if an effector enzyme is activated by some shape feature of the mosaic. The effector enzyme will thus group all input patterns that produce mosaics with the particular shape feature that it recognizes. The number of possible groupings of input patterns depends on the lexicon of different effector enzyme species and their ability to discriminate suitable shape features. The powerful feature of this device is that it converts a symbolic pattern recognition problem into a free energy minimization process. All the clever physics of molecular self-assembly and enzymatic recognition can then be harnessed for transforming input patterns to output activities.

The above (self-assembly) device is a paradigm for cellular information processing. In the case of nerve cells spatiotemporal patterns of transmitter and neurohumoral input are transformed to temporal firing patterns. For real-time processing it is necessary to include a rapid signal propagation capacity in the model. It is likely that the cytoskeleton plays an important role here (Matsumoto & Sakai 1979; Liberman et al. 1985; Hameroff 1987; Conrad 1990). Signals impinging on spatially separated loci on the external membrane would be carried to localized polymacromolecular complexes along cytoskeletal signal lines, where they would then trigger conformational changes in the constituent molecules of the complex. All the clever physics of molecular self-organization would then be brought to bear on neuronal pattern processing in the same manner as in the device pictured in Fig. 6.1, but not in a way that is rate limited by diffusion.

As stated above, the purpose of our gedanken model is to demonstrate in a clear manner that the contribution of proteins (and other bi-

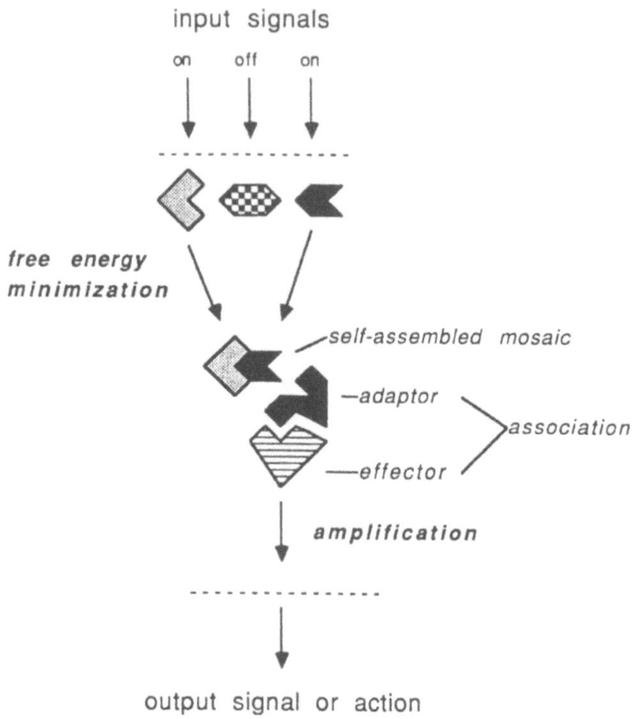

Fig. 6.1. Gedanken device illustrating the role of molecular conformation in cellular pattern processing. Input signals received by receptors at different loci on the cell surface are rerepresented as molecular conformations (shapes) that self-assemble through a free energy minimization process to form a polymacromolecular complex with shape features common to different categories of input patterns. A repertoire of adaptor molecules (or different enzymes with different control sites) links these shape features to the output actions of the cell. The device has nonclassical capabilities since the superposition of electronic states speeds up complex formation, hence enhances the capability of the cell to respond appropriately to different environmental situations.

ological macromolecules) to cellular level pattern recognition cannot be completely captured by modeling them as switches; their ability to act as recognizers per se makes what is probably an even bigger contribution. We could at this point attempt to build up a more realistic picture of the manner in which such conformational processing underlies or interfaces with the biochemical and structural dynamics of the cell. But our argument must travel further down the line, to the quantum level processes that underlie the recognition capability.

First let us consider how remarkable this capability really is. The chance that a key randomly thrown at a lock will dock properly is negligibly small. If the keyhole is enlarged the chance of docking becomes greater, but specific fit is then lost. If the lock is made so that it can dynamically pull the key in the chance of docking would be increased, but the chance of undocking would then become small. Yet enzymes are highly specific, dock rapidly, and are able to undock rapidly.

Conformation-based recognition undoubtedly draws on many physical interactions. But as long as we confine ourselves to classical physical descriptions we face the above conflict between specificity and speed. To escape the classical it is necessary to step down one more scale of size, to the electronic level. In most molecules the effects of electronic motions on nuclear motions can be ignored. The assumption (called the Born–Oppenheimer approximation) is that the electrons follow the motions of the very much heavier nuclei so rapidly that reverse effects can be ignored. This is the case whenever the electrons are tightly bound to the nuclei and at the opposite extreme of conjugated molecules in which the electrons are delocalized in a manner that is incompatible with a constrained three dimensional nuclear structure. Neither condition holds for proteins and the other major biological macromolecules. Here the molecule has definite nuclear shape, and at the same time some of the electrons are only loosely bound to the nuclei; for example, some electrons are now known to tunnel over significant distances in proteins, generally through hydrogen bonds (Beratan et al. 1990). In such cases the interaction between electronic and conformational degrees of freedom becomes important. But electrons are small enough so that the wave properties of matter must be considered. A classical treatment of the time development of a molecular recognition process, such as the formation and decomposition of the enzyme-substrate complex, is for this reason unreasonable.

Very roughly we can picture the situation as follows. Consider two interacting macromolecules, say an enzyme and a substrate. Loosely bound electrons in this system will undergo transitions to nearby energy states due to perturbations by the radiation field or from the nuclear motions themselves. The electronic wave function of the entire system will then be in a superposition of states of different energy. The probability distribution of electrons must then change with time, due to the changing phase relations among the components of the wave function (i. e., due to constructive and destructive interference among the different energy states). The nuclei will be perturbed by the resulting variations in charge distribution. The resulting nuclear motions will then serve as a further source of perturbation to the electronic system. The process may converge to self-consistency (negative feedback between the two systems) or it may self-amplify (positive feedback), finally triggering a significant conformational motion. In the latter, divergent case the particular form of the enzyme-substrate complex will destabilize. The sequence of destabilizations will continue until the electronic and nuclear wave functions become consistent. In the case of an enzyme-substrate interaction the only stable state will be the decomposed state; the form of the complex that allows for the catalytic event is only transient. In the case of self-assembly one of the complexed forms is stable, or more likely, the system cycles among a closely related family of complexed forms.

Here we will not inquire further into the physical details of this process or into the self-consistent field formalism used to treat it (Conrad 1992). The important point is the role of quantum mechanical superposition, that is, of the entirely nonclassical feature that a quantum system is in a set of possible states that interfere with each other. By virtue of the electronic-conformational interaction the superposition of electronic states – the parallelism of the electronic wave function – is harnessed for speed of recognition.

To see what this means at the level of organism function let us look again at the gedanken device illustrated in Fig. 6.1. Recall the essential idea: a symbolic pattern recognition problem is converted to a free energy minimization process. It is only necessary for the molecules interacting to form the mosaic to find a minimum of the energy. Quantum mechanical superposition – roughly something like tunneling – would help them explore the surface. The atomic nuclei are too massive to exhibit significant tunneling. This is not so for the electrons, though, and it is through the interactions of the electrons with the nuclei that the superposition principle is brought to bear on self-assembly and through this on the pattern processing capabilities of the device. Our model of the brain is that it is built from neurons whose pattern processing capabilities are enhanced by intraneuronal molecular recognition processes. The unpicturability of microphysical dynamics is in this way used to enhance the picturable, macroscopic capabilities of organisms.

6.5 The Great Chain of Being Continued

Our travels downward in scale could stop here; but there are two points untouched that should embolden us to travel further. The first is that we have not dealt with wave function collapse, the ability of organisms to convert the nonclassical into the classical, and to make measurements. This has of course been a controversial issue ever since the inception of quantum mechanics. The standard equations of motion, such as Schrödinger's equation, are unitary and entropy conserving. The measurement process is nonunitary, hence irreversible and entropy increasing. We will not here attempt to review the various approaches that have been proposed for reconciling these two different types of time evolution, but simply proceed with the second untouched point, and then with our own proposal for bringing wave function collapse into the picture.

Point two is that molecular recognition entails time evolution. As noted above time evolution occurs whenever a system is in a superposition of states of different energy (cf. Bohm 1951). The classical analogue is acceleration. Thus when an electron is in a superposition of states of different energy it will have some probability of jumping from one state to another. A photon will be emitted and lost to the $1/r$ radiation field – the mark of an accelerated electron. But according to the (general relativistic) principle of equivalence it is not possible to locally distinguish an accelerating frame from a gravitational field. The conceptual awkwardness is that the relativistic equations for the gravitational field are non-linear, therefore incompatible with the superposition principle. If one adheres to the time evolution equations of quantum mechanics it is necessary to accept linearity; if one adheres to the general relativistic theory of gravitation one must in principle introduce nonlinearity. The consequent deviations from the superposition principle would yield an intrinsic wave function collapse.

The suggestion has in fact been made that quantum gravity could play a role in measurement and in yielding up the definiteness of the macroscopic world (Penrose 1989). On the surface deviations from linear superposition deriving from gravity would appear to be too minute to yield any detectable effects. But whether this is the case or not cannot be answered without a model of quantum gravity.

Let us then take the additional step and propose such a model, though without attempting to describe it in any detail here (see Conrad 1991, 1993). The vacuum itself, so-called empty space, is the next lower rung on our ladder of being. We picture this as a plenum of unmanifest fermions, or potentialities for the creation of particle-antiparticle pairs. The Dirac sea of unmanifest vacuum electrons is useful for an initial picture, though ultimately it has to be amended in a number of ways. The observable world, such as the molecules from which organisms are constituted, comprise manifest, positive energy particles. Such particles, to be called absorbers, correspond to lacunae in an unmanifest vacuum sea – essentially ponderable matter is vacuumized vacuum.

The forces between manifest particles are mediated by virtual particle exchanges, as in conventional quantum field theories. However, a virtual particle is now to be identified with a chain of transient fermion-antifermion pairs. For example a virtual photon is a chain of transient electron-positron pairs; a real photon is a chain that carries real energy as viewed from an appropriate co-ordinate system. Gravitons are chains comprising uncharged as well as charged vacuum pairs. Our vacuum sea thus comprises a vast sea of unmanifest vacuum fermions that mediate the gravitational force, and a comparatively very dilute subsea that mediates electromagnetic interactions.

Propagating chains of this sort will be called fluctuons. The basic idea is that electron-positron pairs fluctuate into existence for a very short time, as allowed by the time-energy uncertainty principle. Such pairs must die out immediately in "free space", otherwise conservation of momentum violations would obtain except in a special co-ordinate system. In the neighborhood of an absorber the pair can give up some of its fluctuation momentum in an allowable manner. But it must still decay in a time consistent with the time-energy uncertainty principle. At this point it is in violation of conservation of momentum. Consequently it must decay within a space interval consistent with the position-momentum uncertainty principle. But then it is in violation of conservation of energy. So the pair must be recreated, leading to a violation of conservation of momentum, and so on. The chain propagates until it intersects with a second absorber to which it can transmit its fluctuation momentum.

Trapped fluctuons that do not mediate forces are also possible. Trapping occurs when a transient particle-antiparticle pair in free space annihilates a propagating fluctuon. The vacuum particle will then be trapped in a self-perpetuating fluctuation until it is annihilated through an interaction with another propagating fluctuon. All vacuum particles are thus in reality in superpositions of manifest and unmanifest states. It is in this way that they inherit (or formally acquire) mass and charge. The negative energy assigned to electrons in the Dirac sea is thus a formal property that arises through their interactions in transient positive energy states. (If vacuum particles were purely unmanifest they could never have mass or charge, since they would then never be able to emit or absorb virtual particles.)

Manifest charges repel vacuum electrons, thereby creating a vacuum depression (or polarization) around them. This is a constant feature of all charges, since the overall requirement for charge neutrality prevents asymmetrical build-ups in vacuum density. In the fluctuon model the direct exchange of gravitons is also repulsive. Accordingly manifest masses exert a depressing effect on the full sea of vacuum fermions. Since there is no requirement for overall neutrality, inhomogeneous distributions of vacuum particles will build up in a way that reflects the distribution of positive energy particles. The attractive gravitational interaction between positive energy particles is due to their being pushed together by gravitons emanating from trapped fluctuons in high density regions of the universe. The fluctuon model thus incorporates Mach's principle. It also entails the principle of equivalence, though we will not pursue this point here.

What is most relevant is that the forces between manifest particles depend on the distribution of unmanifest particles, hence on the wave function of the vacuum. The change in state of motion of manifest particles in response to the forces acting on them induces changes in the vacuum density distribution. The forces must be renormalized to reflect these changes. This means that the potentials controlling the time development of the manifest particles depend on the wave function of the vacuum. Thus the proper equations of motion in the fluctuon model should be non-linear with complex potentials (or potentials that must be obtained from a complex function). The linear superposition principle must then fail and the wave function must exhibit some intrinsic collapse as it develops in time (that is, some phase randomization must occur). The time development is thus irreversible and entropy producing (which also follows from the fact that a complex potential is absorptive).

The relation between the manifest and unmanifest structure of the universe is in some respects analogous to the relationship between the electronic and nuclear wave functions of a self-assembling molecular complex. Both must evolve in the direction of mutual self-consistency, and must continue to change if self-consistency cannot be achieved. The difference is that in the treatment of the electronic-conformational interaction the self-consistency approach is only used as a device for approximating a fundamentally reversible equation, but ignoring the heat bath. In the case of the fluctuon model the irreversibility is an irreducible feature.

The universe in this picture is in a sense like a giant homeostat. Deviations from self-consistency produce self-corrective alterations in the vacuum structure. The concomitant irreversibility is negligible except in the presence of very large masses; for ordinary macroscopic objects the entropy would only increase when the system is moving to an equilibrium, since only then would the average relations between manifest and unmanifest vacuum particles be changing. The contribution of this entropy increase to the total entropy increase would be swamped by the "ergodic" component. The effects of the corrective processes (renormalization of the forces) would be averaged over all the absorbers in a random way and the collapse that occurs would at most prevent any microphysical indefiniteness from exhibiting itself at the macrophysical level. The situation is different with biological matter, however, because of the manner in which the choreographed motions of electrons and nuclei are amplified to control the macroscopic structure and behavior of cells and organisms. The homeostatic and information processing capabilities mediated by these intricate electronic-conformational processes draw then not only on the superposition of manifest electronic states, but also on the powerful homeostatic tendencies of the vacuum itself.

6.6 Back to Rules Again

Now let us return to the Turing quintuple representation. On the surface this appears to be unimpeachable, apart from possible limitations on approximation for odd classes of continuous dynamics. After all an organism has states, inputs, and outputs and it has these at every instant of time. Given the state at one instant of time, and the input at that instant, it is reasonable that there is at least a probabilistic rule for the state at the next instant, provided all the relevant information about the organism's past is included in the state. This just means that it is not required that we go back to the distant past to get a better handle on what the next state will be; it should in principle be possible to include all the relevant physical features in the present state description if we work hard enough. The introduction of probabilities slightly generalizes the Turing scheme, but it does not modify the notion of state.

Our quantum analysis of molecular recognition generalizes this picture. We can still talk about an organism's state and rules for the state transitions. But when the organism as a whole is considered, including all its microphysical components, it appears that we must replace the state at any given time by a superposition of possible states (not to be confused with a statistical ensemble). The situation may be described more conveniently by dividing the state of the organism into macroscopic and microscopic variables. The macroscopic variables are classically picturable, or for all practical purposes are classically picturable. Roughly speaking, the microscopic variables must be replaced by collections of possible values that interfere with each other. The picturable aspect of the state, according to our hypothesis, is largely controlled by the nonpicturable aspect associated with the microphysical components.

The fluctuon model further broadens the picture, by attaching wave function collapse to the state transitions. Each time the organism changes, the self-regulatory dynamics of the classically nonpicturable vacuum particles are brought into play. These dynamics couple to the manifest electrons and quasiparticles whose unpicturability enhances the classically picturable capabilities of the organism. The state transitions of the organism (assuming our enriched conception of state) are inherently accompanied by a conversion of possibility into actuality.

But this is just the feature that characterizes everyday rules, particularly of the permissive type. You can cross on WALK. Whether you do or you don't, possibilities are converted to actualities. The subjective experience of decision making, or the formation of an intention, is frequently accompanied by this sense of reduction of possibility into actuality. This is of course a putative psychological correlation. But it is arguably significant that a physical process that appears to be required to satisfy the real time requirements of cognitive behavior has an analogue in the conscious experience associated with the same behavior.

6.7 Of Maps and Territories

Are the capabilities of organisms, then, tied to the particularities of matter? We could consider any of a variety of materials that constitute living systems – carbon, hydrogen, oxygen, nitrogen, water, trace elements – and ask whether maps of the type that enable organisms to exhibit behavior that satisfies the criteria of everyday rules are inseparable from these substances. Let us focus our attention on carbon, since carbon is the pivotal element in biological systems. The valence properties that allow for the formation of stable long chain macromolecules are the main fact. The only other element that comes close is silicon, but the more flexible valence properties of silicon and the heterogeneity of its bonding strengths with other elements (hydrogen, nitrogen, and oxygen) appear to preclude long chains that are stable (cf. Edsell & Wyman 1958). The possibility of significant electronic-conformational interactions, and therefore for harnessing quantum parallelism for molecular pattern recognition, is accordingly greatly reduced. The possibility of coupling the structural and functional capabilities of the organism to the self-regulatory dynamics of the vacuum would also be reduced (assuming that the fluctuon model has merit), since the electronic-conformational interaction is the crucial up-down link for transducing macroscopic state changes to vacuum level disturbances and amplifying the motions that correct these disturbances back to the macroscopic level.

According to our enriched concept of state it is not possible to adequately represent the organism in terms of a sequence of classical states, and therefore not possible to write down a set of Turing quintuples that are isomorphic to its behavior over the whole range of input conditions no matter how large the state set is allowed to be and no matter how complicated each of the inputs is allowed to be. By itself this does not mean that it would be impossible to write down a Turing machine program that could calculate the input-output behavior of the organism, or even calculate the states of the organism. But this program would have to solve the equations of quantum mechanics that describe the interactions among the atoms and molecules in the system, and (if the fluctuon model has merit) the equations that would describe the interactions with unmanifest vacuum particles as well. Let us ignore the fact that we could never ascertain by measurement what these interactions are and that in nature they have to develop through a long evolution process. Let us imagine instead that we simply want to step away from carbon, to silicon for example, and to embed in this alternative physiochemical medium a map – call it an everyday map – that approaches remotely the capabilities of organisms (without here even worrying how this embedding would be achieved). Given the unique properties of carbon for mediating the communication chains that link the strata of possibility to the strata of actuality this is highly unlikely, at least given the facts of chemistry in our universe as they are known today.

Suppose now that we form our silicon into structurally programmable switches and that we take networks of such switches as our alternative physical-chemical media. Assume, for the sake of argument and in accordance with Turing–Church, that such networks have a universal simulation capacity and that one does not have to worry about dynamics that are so sensitive that precise representations of real numbers would be required to compute them. Would it be possible to embed an everyday map in such a system, assuming that time and space bounds are not completely ignored? According to the great chain of being model the map is so tied to the territory that its very existence could even be brought into question by someone inclined toward a constructivist view of mathematics. Moreover, according to the trade-off principle the fraction of interactions that a structurally programmable system can utilize to compute such a map is vastly reduced as compared to the material medium in which the map naturally resides. The possibility of tapping into quantum parallelism and into the self-regulatory dynamics of the vacuum is completely precluded. No interplay between the strata of possibility and the strata of actuality can occur, since the engineer has ensured that all the processing is in the strata of actuality.

Even if the structurally programmable system is allowed an infinite amount of time to do the everyday job it would in principle fail to specify its own interpretation, a capability that must be embedded in the universe at large (Rosen 1991). This is because symbols, and hence sequences of symbols that represent rules, cannot specify their own interpretation (recall from Section 2 that "→" has no intrinsic direction but a material arrow does). Since a map residing on a structurally programmable machine is attached to no material properties it is quite inobvious as to how it could create interpretations (a capability that should clearly be part of the Turing test). Maps that are inseparably tied to the material territory they describe do not suffer this limitation, though as noted above it is no longer clear from the existential point of view that the term "map" has a meaning on its own.

If artificial intelligence and artificial life programs could do the everyday job this would of course discredit the idea that materials-dependent interactions make the main contribution to cognitive capability and that they carry along with them physical processes that have a natural analogue in conscious experience. The awkward question would then arise as to what allows evolution to steadfastly refuse to exploit macromolecular interactions for cognition that it so generically utilizes for morphological development, heredity, immunity, and for all other biological functions. But actual experience with structurally programmable machines has in fact always evidenced strong limitations on domain applicability that justifies treating this question as an argument against the strong claims of AI and against the assumption that everyday maps are separable from the peculiarities of carbon-based chemistry in a watery medium.

6.8 The Three Point Landing

Let us once again reiterate the structure of our argument. Point one is that everyday rules, as simple as they sound, are enormously computationally costly when expressed in terms of the types of rules used to drive either serial or connectionist models of computing. If instead of the humdrum examples used as illustrations we had chosen, say, moral rules the computational requirements would become even more costly. It is conceivable that some aspect, connected with the essence of intentionality in particular and mentality in general, would be missing in principle even in the most elaborate artificial world simulation, even if we allowed this simulation to run to the end of time. But at this stage all we can argue is that this mental aspect is a matter of personal experience. We can not argue in a knockdown way that a structurally programmable machine capable of exhibiting the cognitive behavior of a human doing everyday rules would not enjoy conscious experiences.

Point two is the trade-off principle: in practice it is not possible to implement the cognitive capacities required to implement everyday rules in a structurally programmable machine, and in many cases not possible in principle if it is required that the time scale approach that of everyday behavior. By itself the argument is again not knockdown, for it could always be maintained that time is not important, despite its intimate connection to the quality of our mental experiences. Thus, it might be countered that if the simulation were allowed to run for a very long time – millions of years say – it would be possible to duplicate a minute of human cognitive activity, and that this would in principle be sufficient to endow the simulation with a mental life as well.

Point three is the vertical model of biological information processing – the great cascade of being that connects the macroscopic to the microscopic. This is the domain of computing that, by virtue of being free from structural programmability, escapes the limitations that the trade-off principle would otherwise impose on efficiency and evolutionary adaptability. All the clever physics of molecular specificity is brought to bear on the computational/cognitive capabilities of organisms. The quantum parallelism of the electronic wave function provides for a speedup of these capabilities that could not be accounted for by purely classical models or implementations. Carbon, because of the importance of electronic-conformational interactions, is sine qua non. The dynamics of the vacuum itself, conceived of as a plenum of potentialities, is drawn into the hierarchy of physical-dynamical processes that underlie the capabilities of organisms. Wave function collapse – the continual conversion of the indefiniteness of the microphysical dynamics into the definiteness of macroscopic behavior – is thereby tied to these capabilities.

We cannot prove, by digital simulation or indeed by any traditional mathematical method of a constructive nature, that the great chain of being model provides the computational horsepower required to do the everyday job. This would be a manifest contradiction, since clearly the computational requirements of the simulation would be too great. But this same impossibility supports the hypothesis that the commonplace capabilities of organisms cannot be duplicated by structurally programmable (silicon) machines or by any system not using the electronic-conformational interactions that are so naturally achieved with carbon.

Now we can see how the three points of our argument support each other. Everyday rules demand a lot. The demands can't be satisfied by silicon machines, and probably not by any system not based on carbon molecular chemistry. They certainly cannot be satisfied if they are defined by what carbon molecular systems are capable of doing and in practice could probably not be satisfied by a system of a non carbon nature even if this were structurally nonprogrammable. Along with this material basis comes a physical process – wave function collapse – that is inherent in the nonlinearity of the vacuum dynamics and that is analogous to (and accordingly could naturally underlie) the consciously experienced sense of choice that often accompanies the thought process. Again this leg of the argument by itself is not knockdown, for one might counter with the (ultimately nonfalsifiable) assertion that the whole great chain of being process, including the collapse aspect, is describable by a map that in principle would be realizable by a digital machine with no connection to carbon or the vacuum, if we were willing to ignore all realistic limitations on time and memory resources. Our argument cannot exclude this possibility. But it is to be noted that such a view, to the extent that it admits the phenomenon of consciousness into its ontology, relies on an entirely arbitrary association between the (impossible) physical realizations of certain kinds of mathematical maps and conscious experience. Our interpretation, which chains the great chain of being to itself, provides an objective association between these maps and the physical processes that they describe, and furthermore from the practical point of view it is only when the association is objectively operative that the maps are realizable.

This linkage between the map and the realization – the inseparability property – is the ground on which the three points of our argument converge. As intimated earlier we realize that inseparability raises many questions about the epistemological and ontological status of both the mind-matter and mathematics-matter relation that we have not attempted to fully explore here.

Acknowledgement. M. C. acknowledges partial support for this research from NSF grant ECS-9109860.

References

Beratan, D. N., Onuchic, J. H., Betts, H. N., Bowler, B. E. & Gray, H. B. (1990). "Electron-tunneling pathways in ruthenated proteins", *J. Am. Chem. Soc.* 112: 7915–7921

Bohm, D. (1951). *Quantum Theory*, Prentice-Hall, Englewood Cliffs, NJ

Conrad, D. (1988). *Rule Based Explanations and the Philosophy of Mind*, PhD Dissertation, Wayne State University, Detroit, MI.

Conrad, D. (1993). "Rules, laws, and reductionism", *Appl. Math. Comput.* 56: 341–358

Conrad, M. (1985). "On design principles for a molecular computer", *Commun. ACM* 28: 464–480

Conrad, M. (1988). "The price of programmability" in *The Universal Turing Machine: A Fifty Year Survey*, ed. by R. Herken, Oxford Univ. Press, Oxford, UK, pp. 285–307

Conrad, M. (1989). "The brain-machine disanalogy", *BioSystems* 22: 197–213

Conrad, M. (1991). "Transient excitations of the Dirac vacuum as a mechanism of virtual particle exchange", *Phys. Lett. A* 152: 245–250

Conrad, M. (1993). "The fluctuon model of force, life, and computation: a constructive analysis", *Appl. Math. Comput.* 56: 203–259

Conrad, M. (1987). "Molecular computing: a synthetic approach to brain theory" in *Real Brains, Artificial Minds*, ed. by J. Casti & A. Karlqvist, North Holland, New York, pp. 197–226

Conrad, M. (1990). "Molecular computing", in *Advances in Computers*, vol. 31, ed. by M. C. Yovits, Academic Press, San Diego, CA, pp. 235–324
Conrad, M. (1992). "Quantum molecular computing: the self-assembly model", *Int. J. Quant. Chem.: Quantum Biology Symposium* 19: 125–143
Conrad, M., Kampfner, R. R., Kirby, K. G., Rizki, E. N., Schleis, G., Smalz, R. & Trenary, R. (1989). "Towards an artificial brain", *BioSystems* 23: 175–218
Conrad, M. & Rizki, M. M. (1989). "The artificial worlds approach to emergent evolution", *BioSystems* 23: 247–260
Edsall, J. T. & Wyman, J. (1958). *Biophysical Chemistry*, Academic Press, New York
Grice, H. P. (1975). "Logic and conversation", in *Syntax and Semantics 3: Speech Acts*, ed. by P. Cole & J. L. Morgan, Academic Press, New York, pp. 41–58
Hameroff, S. (1987). *Ultimate Computing*, North Holland, Amsterdam
Haugeland, J. (1981). "Semantic engines, an introduction to mind design" in *Mind Design*, ed. by J. Haugeland, MIT Press, Cambridge, Mass, pp. 1–34
Levinson, S. (1983). *Pragmatics*, Cambridge University Press, Cambridge
Liberman, E. A., Minina, S. V., Mjakotina, O. L., Shklovsky-Kordy, N. E. & Conrad, M. (1985). "Neuron generator potentials evoked by intracellular injection of cyclic nucleotides and mechanical distension", *Brain Res.* 338: 33–44
Matsumoto, G. & Sakai, H. (1979). "Microtubules inside the plasma membrane of squid giant axons and their possible physiological function", *J. Membrane Biol.* 50: 1–14
Penrose, R. (1989). *The Emperor's New Mind*, Oxford University Press, Oxford
Rosen, R. (1991). *Life Itself*, Columbia University Press, New York
Searle, J. (1984). *Minds, Brains, and Science*, Harvard University Press, Cambridge, MA

Part III

Network and Systems Level

A common theme throughout this book is the relationship between microscopic and macroscopic levels in neural systems. In the previous section we were given concrete examples on how cognitive functions could be rooted in sub-cellular structures and processes. At such a microscopic level, quantum mechanical effects might play a role, as was also suggested by Penrose in the first block. However, so far, there is no evidence that any one organizational level or part of the brain would hold the key to the phenomenon of mind. Subsystems and sublevels should always be put in a larger context, as was pointed out by Rosen in the first block. Traditionally, this has not been done in science. Nature has instead been regarded, particularly by physics, as primarily consisting of simple systems, systems that in principle were mathematically describable. It is only recently that physics has begun to recognize and study more complex systems. This has partly been made possible through the development and usage of powerful digital computers. Indeed, the last decade has seen a tremendous development and usage of computational models of complex dynamical systems, based on theories in physics but hitherto undescribable by analytical methods. In particular, this is the case for the research field of artificial neural networks (ANN), which has led to the so-called connectionist view of the brain and its functions.

While there is nothing that points at any particular level of description that should be crucial for the mind-matter problem, scientists from different fields and backgrounds turn their interest to various levels. Haken and Liljenström, in the two following chapters, have chosen to model brain functions at the level of neural networks. Here, events and structures at the cellular or sub-cellular levels are not directly included. Indeed, events at those levels would normally be too noisy and irregular to be useful for the neural computations presumed to underlie cognitive functions, according to a connectionist view. (The same idea was also expressed by Schrödinger in his book "What is Life?", 1944). Cognitive processes would require the co-operative action of a large number of neurons or neuronal events, and quantum effects would be negligible in comparison with the mass action of "macroscopic" neurons, interconnected in large networks. Clearly, this view deviates from that presented in the first two blocks of this book.

It would probably be an oversimplification to assume that all aspects of mentality were based on the same mechanisms or could be described with the same theoretical framework. In this book it is debated to what extent mental phenomena are rooted in computational or non-computational processes. For example, as we have seen earlier, Penrose and Hameroff argue that the non-computational component is essential, whereas perhaps the dominant view is that all biological (and mental) processes are computational (Dennett 1991). Yet, the question may not be an all-or-nothing issue. Could perhaps some aspects of mind, such as sensory perception, classification, learning, and recall, be based on neural computations that would be rather stable and "deterministic"? Would such processes be describable at a macroscopic, "classical" level, whereas other aspects, such as consciousness, understanding, or qualia possibly could be of a different nature? The "connectionist" approach, as exemplified by the two chapters in this block, focuses on those aspects of mental activity that are thought to be computational and describable with the techniques and level of understanding we have today. This could even imply studying aspects such as visual awareness (cf. Crick & Koch 1991), although usually much less ambitious objectives are at hand. With a few exceptions (e. g., Dennett 1991), the problem of subjectivity is not addressed in a typical connectionist approach.

With this kind of approach, brain structures are described in physical-mathematical terms as complex (non-linear) dynamical systems, and cognitive functions are regarded as an emergent property of such systems. Historically, the brain has been described as the most advanced technical device or paradigm of the period. For example, a hydraulic engine, a telephone switchboard, and a digital computer have been used as metaphors for the brain and its functioning. With the current metaphor, one implicitly assumes that certain brain functions, including some cognitive functions, can be adequately described using computational methods and mathematical analysis of complex systems.

A complex system is not just a complicated system, in fact, it might be based on a foundation of very simple mathematical rules. The degree of complexity can be defined as the minimum amount of information, or number of algorithmic steps (lines of a computer program), required to describe any particular system or process. A complex system is often characterized by having a large number of components, or degrees of freedom, by being open and non-linear. Its behavior is typically discontinuous and unpredictable. As the degree of complexity increases, new and unpredictable properties and functions may appear at each new "level" of complexity, or organization. An example of such emergent properties is the sound waves that suddenly can appear in a container when the number of gas molecules is increased (Hopfield 1986). One aspect of evolution is that it brings about organisms with increasing complexity as regard to, for example, genetic organization and behavior. In particular, the nervous system has become more and more complex, enabling new and more elaborate functions to appear. A primitive nervous system can allow for simple sensation and motor action, whereas a more developed one could include perception, learning, and associative memory. With the organization into several neocortical areas, the nervous system may allow for cognition, planning, and higher aspects of consciousness. How does this picture fit with a view where quantum effects, that presumably have been present throughout the history of life, would be crucial for cognition and consciousness? Do these different views exclude each other, or could there be yet undiscovered principles that link the two?

Treating the brain as a complex physical system, is it possible to use modeling techniques from physics to describe aspects of the neural dynamics that could be relevant for certain brain functions? Could computer simulations with carefully developed models point at what structures and what parameters are related to a certain behavior? A common view is that this indeed is possible. Inevitably, though, with such an approach one has to make a rather far-going abstraction of the brain's structure and organization, an abstraction that is necessary for a mathematical treatment. Simplification is not only by necessity, but when attempting to understand the behavior or function of a complex system like the brain, one must extract those system properties that supposedly are relevant. The crucial questions: Which details are important and which can be excluded in the description? The strength of a model is determined by its ability to describe (an aspect of) the real system with as few parameters as possible. With a powerful description one should be able, at least in principle, to predict the (gross) behavior of the system. Most models consider networks of more or less simple processing units, corresponding to single neurons (e. g., Wilson & Bower 1992), or more commonly, to large populations of neurons (Hopfield 1984). The biological realism also differs with regard to network structure.

Haken applies his theory of self-organizing systems, synergetics (Haken 1983), to describe and model how the brain processes, in particular, visual signals and performs pattern recognition (Haken 1991). The success of this method is clearly shown by the examples with computer recognition of human faces, even when the input is partial, translated, rotated, or noisy with respect to the learnt input patterns. The order parameter concept of synergetics (Haken 1983) is shown to be a useful tool in modeling the behavior of co-operating neurons at an adequate level. By extending the theory, could thoughts be interpreted to be such "order parameters" that "enslave" the neurons of the brain? If so, there would be a two-way interaction, a bottom-up influence from neurons to cognitive functions (thoughts), but also a top-down one from thoughts to neurons. An attention parameter could also change the perception of a complex scene and, for example, describe how the oscillatory perception of ambiguous figures might occur in the brain. The synergetics approach is further used for analyzing EEG and MEG and relate such electrical activity to the dynamics of the neural networks.

As discussed above, it is important to understand the various brain functions from an evolutionary point of view. This point is stressed in the chapter by Liljenström. The neural system and its dynamics is here viewed as having evolved to cope efficiently with rapid changes in a complex environment. Is there perhaps an evolutionary pressure to maximize the rate of information processing, as well as the storage capacity of the nervous system? Could the network structure and complex dynamics of the brain, including oscillations and chaos-like behavior, serve this purpose? These ideas are tested with a neural network model of the olfactory cortex, which mimics the characteristic dynamics and presumed memory functions of the real system. Input patterns are stored in the network connections as oscillatory (limit cycle attractor) states. It is shown that this dynamics can be used for reducing the recall time of any specific stored pattern in associative memory tasks. In addition, neuronal noise is shown to improve performance under certain circumstances and optimal noise levels can be found for which the recall time is minimized. Noise and chaos are also useful for transitions between various memory states, presumably important for a continuous thought process. The dynamics can be under neuromodulatory control, shifting the balance between flexibility (sensitivity) and stability, so that the system gives an appropriate response to any internal or environmental change.

The chapters by Haken and Liljenström both give a similar view of the brain as a complex dynamical system, using model examples of two different sensory systems, vision and olfaction (smell). This approach seems to be specifically useful for pattern recognition, presumably similar to that going on in our brains. It thus gives us a way of relating the neurodynamics to some basic cognitive functions, such as (some aspects of) perception, learning, and associative memory. Although both chapters give a rather optimistic view of the possibilities of describing some brain functions mathematically, and simulating them on computers, it is also acknowledged that the models can only capture a minute part of the brain's immense complexity. In particular, when it concerns the relationship between neural events and the subjective experience, we may never get a satisfactory scientific description, at least not with the science of today. This view was strongly held by Penrose and Rosen in the first block, but contrasts that of Dennett (1991), where the whole issue of consciousness is regarded as an already solved problem. The problem of consciousness and subjectivity is further discussed in the fourth and last block that follows.

References

Crick, F. & Koch, C. (1990). "Towards a neurobiological theory of consciousness", *Seminars in the Neurosciences*. Vol. 2, pp. 263–275.

Dennett, D. C. (1991). Consciousness Explained, Little, Brown and Company, Boston, MA

Haken, H. (1983). *Synergetics: An Introduction*, Springer-Verlag, New York

Haken, H. (1991). *Synergetic Computers and Cogniton*, Springer-Verlag, Berlin

Hopfield, J. (1984). "Neurons with graded response hace collective computational properties like those of two-state neurons". *Proc. Natl. Acad. Sci. USA* 81: 3088–3092

Hopfield, J. (1986). "Neural networks and physical systems with emergent computational abilities" in *The Lesson of Quantum Theory*, ed. by J. de Boer, E. Dal & O. Ulfbeck, North-Holland, Elsevier, Amsterdam

Wilson, M. A. & Bower, J. M. (1989). "The simulations of large-scale neural networks" in *Methods of Neuronal Modeling: From Synapses to Networks*, ed. by C. Koch & I. Segev, MIT Press, Cambridge

Schrödinger, E. (1944). *What is Life?* Cambridge University Press, Cambridge

7. Synergetics of the Brain

Hermann Haken

7.1 Introduction

The human brain is certainly the most complex system we know of. It has a complex structure composed of about one hundred billion neurons that are connected with each other in a complicated fashion. Just to visualize this number, let us enlarge the neurons so that one hundred of them can be put into a thimble with the volume of one cubic centimeter. Then we need a house of ten meters length, ten meters depth and ten meters height to accommodate these thimbles in a close packing.

The brain exhibits complex behavior that becomes manifest either by behavior, such as movements, speech, etc., or by direct measurements of its electro-chemical processes. Thus the brain shows very many facets and its study is in a way inexhaustible. So we are confronted with what are the sensible questions to ask about the brain.

As it will transpire from the articles in this volume, the questions we ask may depend on our taste, background, the available experimental techniques, and theoretical concepts and mathematical tools.

7.2 The Brain as a Black Box

According to behaviorism, in particular according to Skinner, we do not worry about the interior of the black box *brain*, but rather focus our attention on its behavior. In this approach, one studies the responses of a test animal or a human being to specific signals. In many cases this approach has been more or less abandoned when the behavior of humans is studied, because humans can communicate about their introspection. Different fields of scientific study are concerned with the corresponding investigations, such as psychology with respect to vision, hearing and behavior, or psychiatry in the case of mental illnesses, etc.

7.3 Structure of the Brain

The brain is composed of individual building blocks, the neurons. Golgi was the first to make them visible by staining. There are about twenty

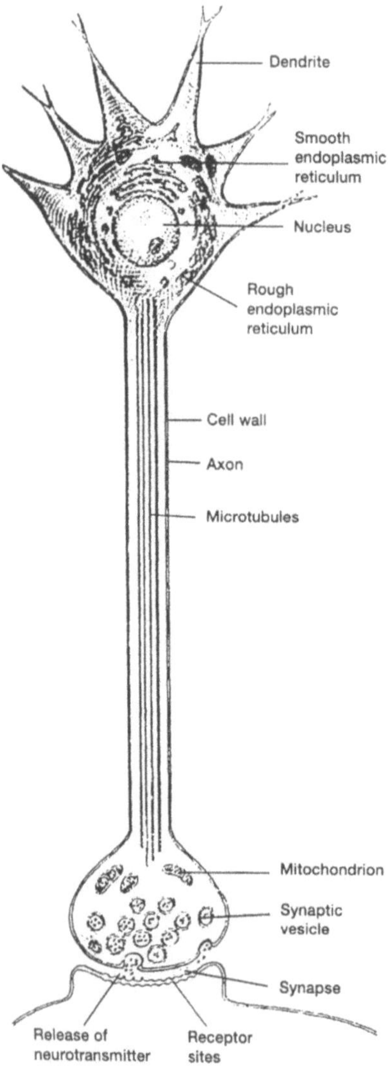

Fig. 7.1. A neuron with its axon and dendrites

different kinds of neurons, such as pyramidal cells, Purkinje cells, etc. Each neuron possesses an axon through which it sends its signals in form of nerve pulses (Fig. 7.1). On the other hand, it collects signals from other cells via its dendrites. The places of contact between cells are the synapses. When a synapse is excited by an electric impulse, it releases chemical substances, the so-called neural transmitters, so that the signal is transferred through the synaptic cleft to the other side belonging to another nerve cell. According to our present knowledge, a neuron collects incoming nerve impulses, sums them up, and sends out a new impulse to other cells provided the sum of the incoming signals exceeds a special threshold. The intensity of the outgoing signal is pulse-code modulated, i. e., the neuron emits pulses of equal length but varying distances, where the distance between the pulses becomes the smaller the higher the intensity of the signal (Fig. 7.2).

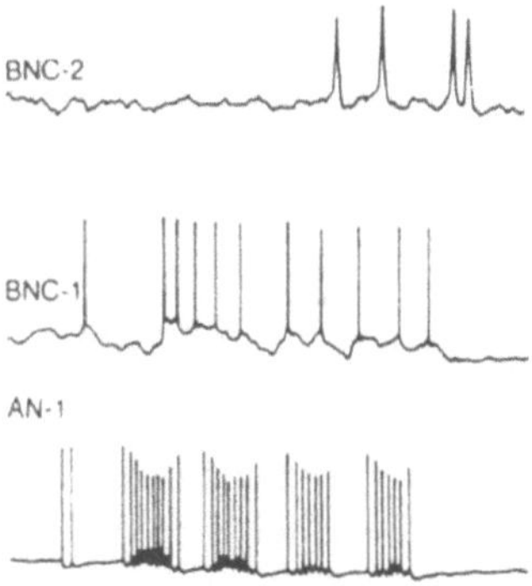

Fig. 7.2. Examples of nerve pulses. Voltage is plotted against time

7.4 Structure and Function: Macroscopic

Roughly speaking, different regions of the brain are devoted to different tasks (Fig. 7.3). The localization of functions can be identified by lesions. For instance, when people suffer from strokes in their left brain hemisphere, specific centers may be damaged so that speech is impeded. The Brocca center is responsible for grammar, whereas the Wernicke center is responsible for contents. In the Russian-Japanese war, the Russian army used a new rifle whose high-speed projectiles produced lesions in

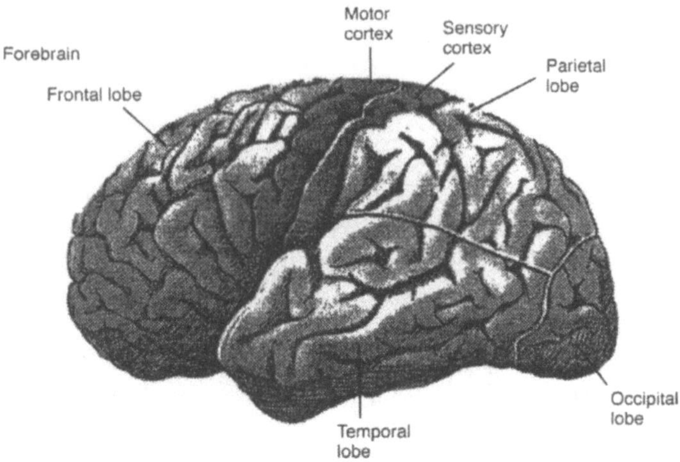

Fig. 7.3. The brain and some of its important parts

the rear part of the brain without damaging, for instance, the eyes. Nevertheless, soldiers became blind and the Japanese medical doctor Inoui showed that vision is processed in what is now called *the visual cortex*. In order to treat patients suffering from epilepsy, the so-called *corpus callosum* that connects the left and right brain hemispheres was cut. As Sperry found in ingenious experiments, both hemispheres fulfill, roughly speaking, different tasks. For instance, the left hemisphere is more responsible for sequential and rational tasks, whereas the right hemisphere processes preferably recognition of scenes, music, etc. In our modern times, X-ray tomography and magnetic resonance, formerly called nuclear magnetic resonance, play an important role in finding changes of the brain tissue, for instance because of tumors. PET-scans (positron emission tomography) allow one to study activity in the following way: When brain centers are fulfilling specific tasks, such as reading, their local metabolism is enhanced. This metabolism can be measured by the concentration of glucose. If glucose is made radio active so that it can emit positrons, the positrons can be registered and in this way the concentration of glucose measured (Fig. 7.4). This method becomes highly sensitive when differences between glucose concentrations with and without brain activity are measured.

A further method of studying brain activity is by electro-encephalogram and magneto-encephalogram. In the first method, electrodes are

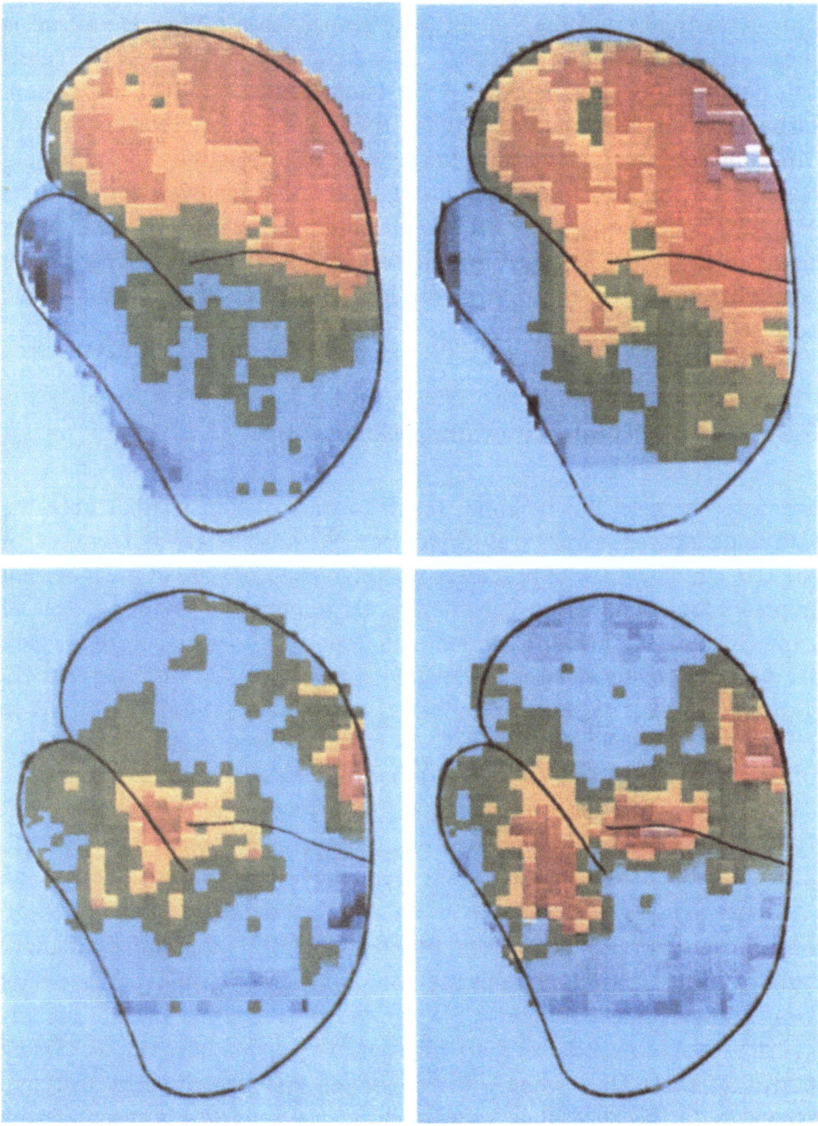

Fig. 7.4. Example of PET-scan. The yellow and red regions show enhanced activity of the brain under different functions

put on the scalp and the voltage with respect to a reference voltage is measured as a function of time. Using frequency filters, specific bands can be identified, for instance the α-band around the 10 hertz region. Depending on the brain's activity, the electric brain waves show a different behavior, for instance the α-waves show activity in the resting state and break down when the test person opens his eyes, for instance. Magneto-encephalograms give a rather high resolution of the magnetic fields in the brain. By the EEG different kinds of brain activity can be easily distinguished, for instance different phases of sleep or an epileptic seisure.

7.5 Structure and Function: Microscopic

We refer here to the neuronal level as microscopic so that here the individual neuron is treated as a black box. We must be aware, however, of the fact that the study of processes within the neurons are of great current interest, for instance the behavior of its microtubuli. A good deal of experiments have been performed on neurons of the visual cortex, where the neurons are arranged in different layers or areas, for instance area 18. The now-famous experiment by Hubel and Wiesel (1965, 1977) runs as follows: An electric signal is derived from a neuron in the visual cortex of an anesthesized cat. To this cat a bar moving in a specific direction is shown. Then one may, for instance, identify orientation sensitive cells. If the bar is oriented in one direction, there is no response from the cell. The cell responses measured by the frequency of the pulses increase with increasing angle, reaches a maximum, and then decrease again (Fig. 7.5). More recently, it has been found by Freeman (1991) that in the olfactory bulb of some animals, neurons may fire in synchrony. In experiments by Gray, Singer, and co-workers (1990) and by Eckhorn, and co-workers (1990), synchronized firing of neurons was found also in the visual cortex of cats. Thus when a bar is moved and perceived by different receptors in the eyes, their corresponding neurons in the visual cortex can fire in synchrony even if the neurons are a few millimeters apart. The physiological meaning of these synchronization effects, which last only for a few pulses and then decay, is not clear at the present moment. These synchronization effects may serve for a background-foreground discrimination or may have a general role in binding features together as has been speculated for instance by Singer.

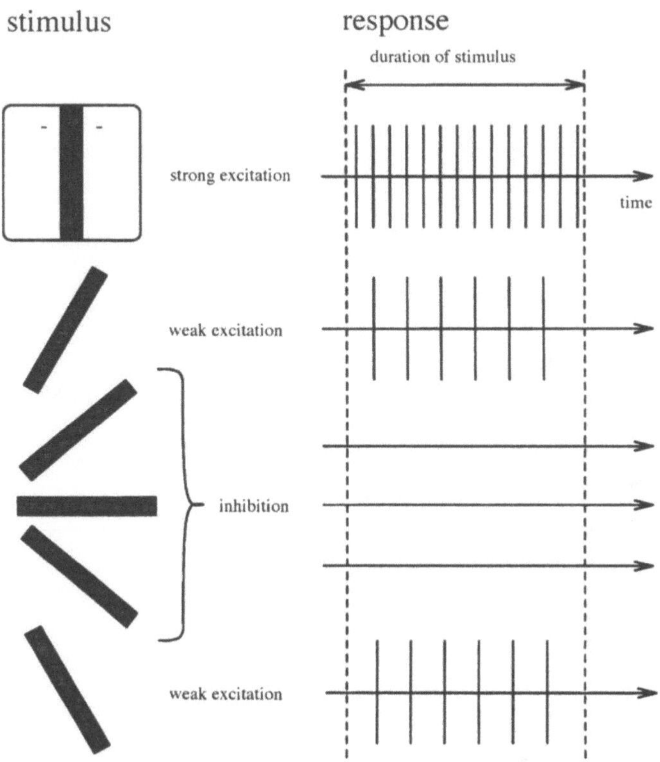

Fig. 7.5. Response of a cell of the visual cortex to a bar shown to a cat. Note that the pulse frequency changes drastically as a function of the angle of the bar with respect to the horizontal axes

7.6 Learning and Memory

From the experimental point of view, the localization of learning and memory is not clear at the present time. According to a famous hypothesis by Hebb (1949), learning takes place by reinforcement or weakening of synapses. When two cells connected by a synapse are simultaneously strongly active, this synapse will be enhanced, otherwise weakened. Detailed studies of learning in the sense of habituation or sensitization were performed by Kandel (1979) on the sea snale *aplysia*, where complicated chemical transformations could be found. These processes can explain short- or medium-time memory but not long-time memory.

7.7 Theory: Artificial Intelligence

When programmable computers came into existence via the work of Konrad Zuse, John von Neumann, and others, and Turing's concept of the universal Turing machine became known, it was tempting to compare the action of the brain with that of a computer. This was even more so when it became clear that computers can process not only numbers but also symbols. In the frame of artificial intelligence, thinking was conceived as a serial processing of symbols. However, it is noteworthy that in 1943 McCulloch and Pitts (1943) conceived the brain as a parallel network, and showed that such a network can perfom all the logical operations of Boolean algebra. The network was composed of model neurons, each of which possesses two states, one resting state and one active state, during which a signal is emitted. The neurons are interconnected. If the sum of the inputs received by one neuron remains below a threshold, the neuron remains at rest, otherwise it emits a sigal. A simple network of this kind was realized by Rosenblatt (1958a, 1958b), who called it *perceptron*. Later on, Hopfield (1982) showed that such a network can be realized by a spin glass in solid state physics. Of conceptual importance is the fact that via this analogy Hopfield introduced the concept of an energy surface into neural networks.

7.8 The Synergetic Approach to the Brain

As mentioned in the introduction, the brain is a complex system. It is also a dynamical system because of its many electro-chemical processes underlying its functioning. In my opinion it is above all a synergetic system. Indeed, in synergetics we study the relationship between the microscopic and the macroscopic level of a system (Haken 1983, 1993). The microscopic level is characterized by the individual components of a system; through their interactions these components may produce qualitatively new features at the macroscopic level. This kind of emergence of new qualities is ubiquitous in the brain. By the cooperation of many cells, perception, motor-action, thinking, speech-control, etc., are brought about. Incidentally, the synergetic point of view introduces new concepts as is shown Table 1.

In it, on the left-hand side, the synergetic concepts are shown, whereas the right-hand side contains the more traditional ideas. According to the

Table 1.

synergetics	conventional
neural network	individual cell ("grandmother-cell")
delocalized	localized
self-organized	computer (programmable)
parallel	serial
unstable	stable

traditional concept, recognition of a face is eventually performed by an individual cell, which was called the grandmother cell, because it may recognize one's grandmother. According to synergetics, perception is based on the action of a whole ensemble of cells at a macroscopic level. In contrast to conceiving the brain as a computer that may be programmed, we conceive the brain as a self-organizing system close to instability points. In order to make the concepts of synergetics available for a theory of the brain, we briefly remind the reader of some basic concepts of synergetics.

7.9 Basic Concepts of Synergetics

We consider a network composed of many components that we distinguish by an index $j = 1, \ldots n$. The variable q_j describes the activity of component j, where q_j may not be only one variable but may also comprise several variables to characterize the activity of that component. The components are lumped together into the so-called state vector $q = (q_1, q_2, \ldots)$. To study the behavior of the system, we are interested in the evolution of the system in time, i. e., in the temporal change of q. This change is determined by the present state, by so-called control parameters, and by chance events. Let us elucidate the role of the control parameter. A well-known example in the field of physics is that of a fluid heated from below. By the heating a temperature difference between the lower and upper surface is generated. This temperature difference $\alpha = T_1 - T_2$ is a so-called control parameter. When α is small, the fluid remains at rest, but when α exceeds a critical value, the fluid starts a specific ordered motion, for instance in the form of rolls (Fig. 7.6). Thus the behavior of the fluid is controlled by the control parameter α. On the other hand, the control parameter does not anticipate the pattern forma-

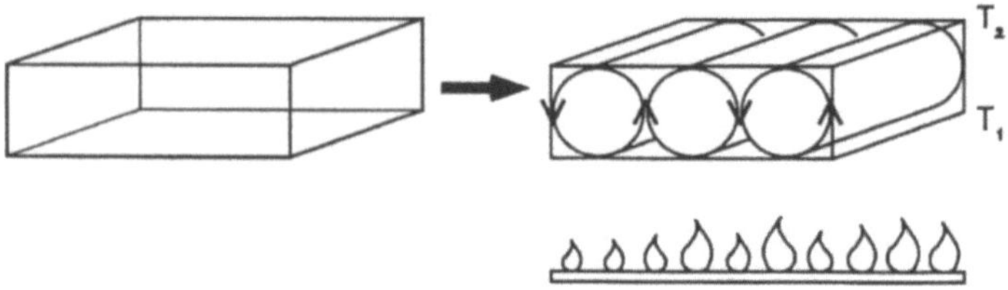

Fig. 7.6. Formation of rolls in a fluid

tion, because the heating is entirely uniform and does not directly cause the inhomogeneous movement. Rather this specific kind of movement is brought about by self-organization. Control parameters applying to the brain may be concentrations of neural transmitters, such as Dopamin or Serotonin, or drugs administerd from the outside, such as Haloperidol or hormones. As we shall elucidate below, microscopic chance events may play an important role by triggering macroscopic processes. In view of the high complexity of the brain but also of other systems, at first sight it might seem to be impossible to cope with the problem of the emergence of new qualities at macroscopic scales. However, the following strategy has proved extremely useful. Let us consider the still very simple example of the fluid heated from below. For a given control parameter value α_0, the state of the system q_0 is known. In the present case, it is the resting state. When we change $\alpha_0 \rightarrow \alpha$, suddenly a new state q may evolve; for instance, the roll pattern may emerge. As is shown quite generally in synergetics, this macroscopic motion can be described by one or several order parameters ξ. According to the slaving principle of synergetics, the order parameters ξ determine the behavior of all the individual parts of the system, for instance the motion of the individual volume elements. The number of components may be very large, whereas, in general, the number of order parameters is rather small or may even be only one. In this way the slaving principle implies an enormous information compression. Instead of characterizing the system by its many individual components and their activities, it is sufficient to characterize it by the order parameters. It must be noted, however, that the order parameters are not quite like puppet players acting on the puppets. In fact, as is shown in synergetics, the individual parts of the system react on the order parameters. In this way, we observe the phenomenon of circular causality. The order parameters determine the behavior of the individual parts, but on the other hand, the individual parts fix the behavior of the order parameters. An important difference between order parameters and enslaved parts consists in their time-scales. Order parameters change on a slower time-scale than the enslaved parts. Because the typical time-scale for the action of a neuron is a millisecond, while that of brain activities, such as perception, motor-control, etc., is a hundred milliseconds, there is a clear-cut separation in time-scales which encourages us to apply the concepts of synergetics to the brain. Because of the slaving principle, we may study the macroscopic events directly by means of the order parameter concept.

Let us consider in the following sections a few typical cases of order parameters that can be identified also in brain activities.

7.10 Dynamics of One Order Parameter

As is shown in synergetics, the behavior, i. e., the time-dependence, of a single order parameter can be visualized as follows: When we plot the order parameter along the abscissa of Fig. 7.7, we may identify its time-dependence with that of a ball moving in a hilly landscape as depicted in that figure. If the order parameter is disturbed, it will relax towards its equilibrium point. While this behavior is true for a certain range of control parameters α_0, at a critical control parameter value $\alpha = \alpha_1$ the bottom of the valley becomes very flat (Fig. 7.8). When we imagine that the order parameter is still subject to small chance events, called fluctuations, we realize that the ball will be kicked like a ball by soccer players with rather large elongations on the flat bottom. That means that the fluctuations of the order parameter become rather large – we are dealing with *critical fluctuations*. Furthermore, it takes the ball a rather long time to reach its equilibrium point, a phenomenon called *critical slowing down*. When finally the control parameter value exceeds its critical value α_1, the potential landscape is deformed into that of Fig. 7.9. The position of ξ in the middle of the landscape has become unstable and is replaced by two stable positions. Here we are dealing with the phenomenon of so-called bistability. Because the system can go only into the left or the right minimum, it has to break the symmetry. The transition from Fig. 7.7 to Fig. 7.9 is called a nonequilibrium phase transition, well known in such physical systems as fluids or lasers. Such bistability occurs also in human perception. For instance, in Fig. 7.10 a physicist may recognize a microscopic picture of a ferromagnet, another person will recognize a stock market. In other systems, the order parameter dynamics may be

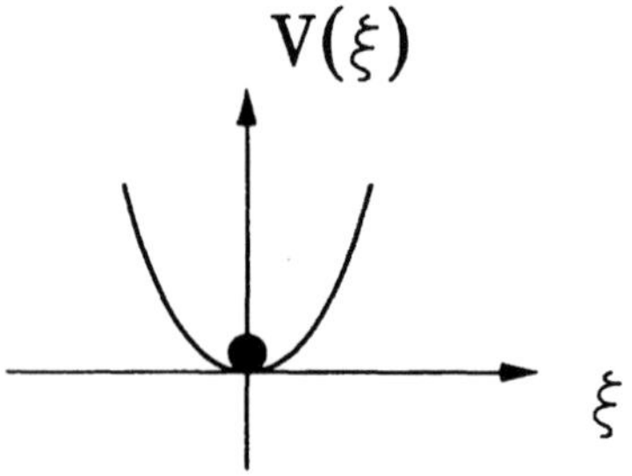

Fig. 7.7. The potential landscape in which potential function V is plotted against the order parameter ξ for a control parameter value below its critical value

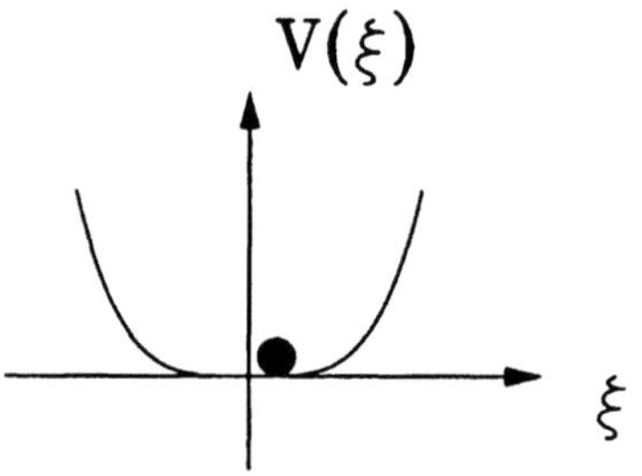

Fig. 7.8. Same as Fig. 7.7, but for a control parameter value at the critical point

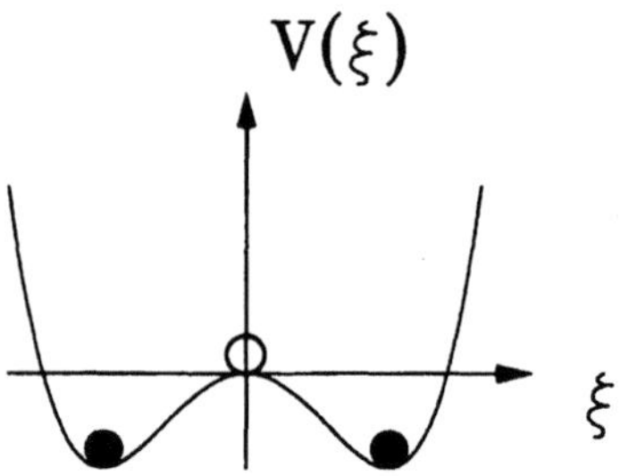

Fig. 7.9. Same as Fig. 7.7, but for a control parameter value above its critical point

described as motion of a ball in a landscape such as that of Fig. 7.11. When a control parameter is changed, the landscape may be deformed as shown in that figure, starting from the left to the right. In this case, the ball will fall down to the second minimum at the control parameter value α_2. On the other hand, when we follow the change of the control parameter in the reverse direction, quite obviously the ball will acquire its original position only under the control parameter value α_0. In other words, for the same

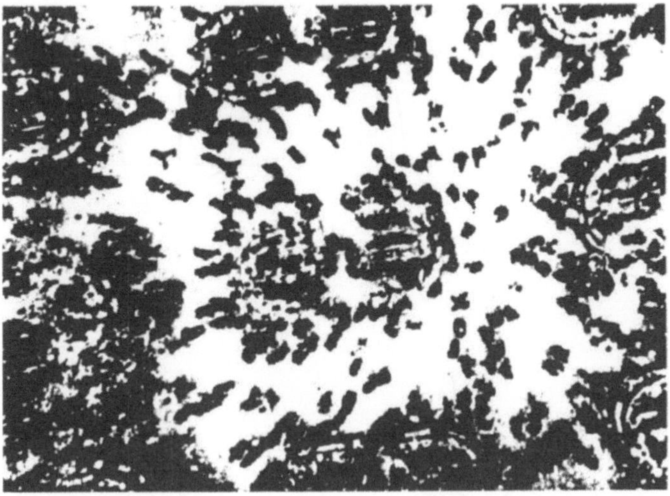

Fig. 7.10. Example of bistability in perception Ferromagnet or stock market? (after Stadler & Kruse 1990)

control parameter value α_1 there are two states possible depending on the history of the system, a phenomenon called *hysteresis*. Such hysteresis phenomena can also be observed in perception as is shown in Fig. 7.12.

7.11 Several Order Parameters

If there are several order parameters, in a number of cases their behavior can again be described as that of a ball in a landscape. For instance, if there are two order parameters, a landscape may look like a golf course with its individual holes. However, in the case of two order parameters also oscillations may occur. Such oscillations can be observed also in the case of perception of ambiguous figures as is exemplified by Fig. 7.13. Here one may recognize at first either an old woman or a young woman, but then the first percept fades away, gives way to the other one, which again fades away, giving way to the first one, etc.

Finally, in the case of three or more order parameters, in addition to stable states and oscillations, also deterministic chaos can occur. In a fundamental paper Agnes Babloyantz (1985) showed that in certain stages of sleep, chaotic states of the EEG of low dimensions, about five or six, could be identified. This finding lends strong support to the idea that the order parameter concept can be applied to a number of brain functions.

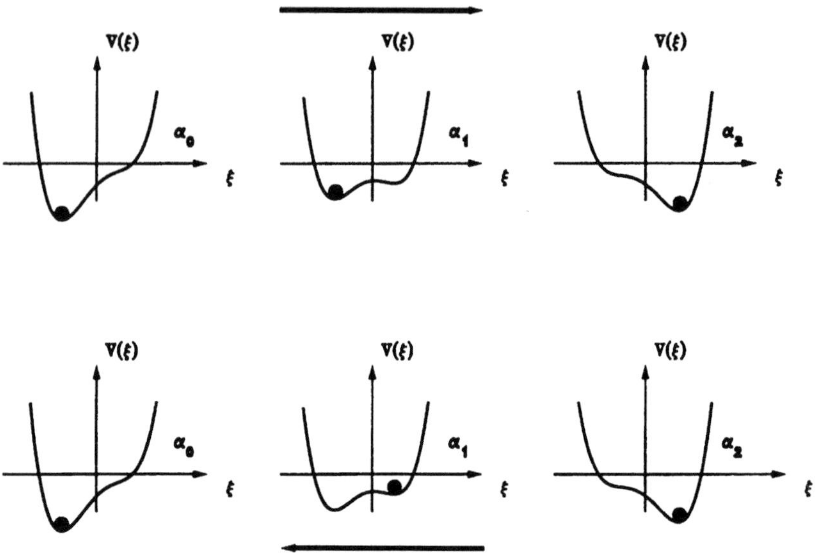

Fig. 7.11. Example of hysteresis. Deformation of a landscape with the changing control parameter. Upper row from left to right: The ball remains at its original position and drops to the other minimum only at the final last stage. Lower row from right to left: The ball stays in its initial position now and returns to its original position only under the same control parameter value with which the upper row started

In the following I wish to sketch some more detailed models of brain functions in terms of synergetics. The models concern pattern recognition, recognition of ambivalent patterns, and EEG and MEG analysis.

7.12 Pattern Recognition

So far we have given a qualitative description of synergetics. In order to show how these concepts can be put into an operational form, we consider the problem of pattern recognition (Haken 1991) by humans and machines. Our approach to pattern recognition is based on three ingredients: First, and in accordance with the generally accepted belief, we identify pattern recognition with the action of an associative memory (Kohonen 1987). An example for an associative memory is provided by a telephone book. For instance when we look up the name Adam Miller, the telephone book tells us, in addition to the name, his telephone number.

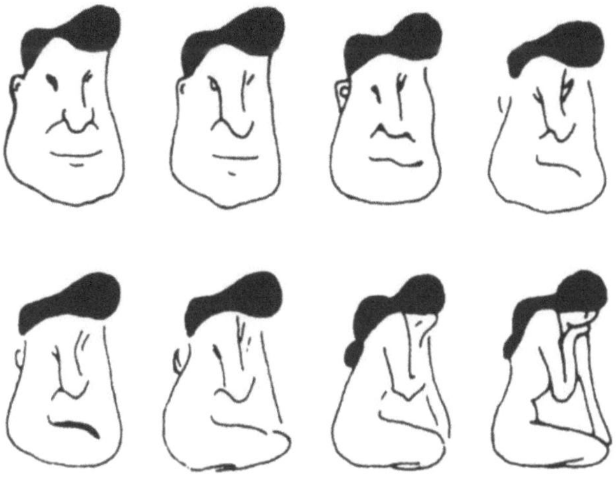

Fig. 7.12. Example of hysteresis in visual perception. The switch from face to woman and back depends on the previously perceived patterns

Fig. 7.13. Example of an ambiguous pattern: young woman or old woman?

Thus associative memory serves the completion of a set of data. The second ingredient of our approach is the assumption that the associative memory is realized by means of a dynamics of order parameters within a potential landscape in analogy to a landscape of Fig. 7.9, where, however, we have now to deal with a multistable landscape. The third and most essential ingredient of our approach is the idea that pattern recognition is nothing but pattern formation. To this end let us consider a simple example from fluid dynamics (compare Fig. 7.14). Here we simulate the behavior of a liquid that is heated from below so that roll patterns

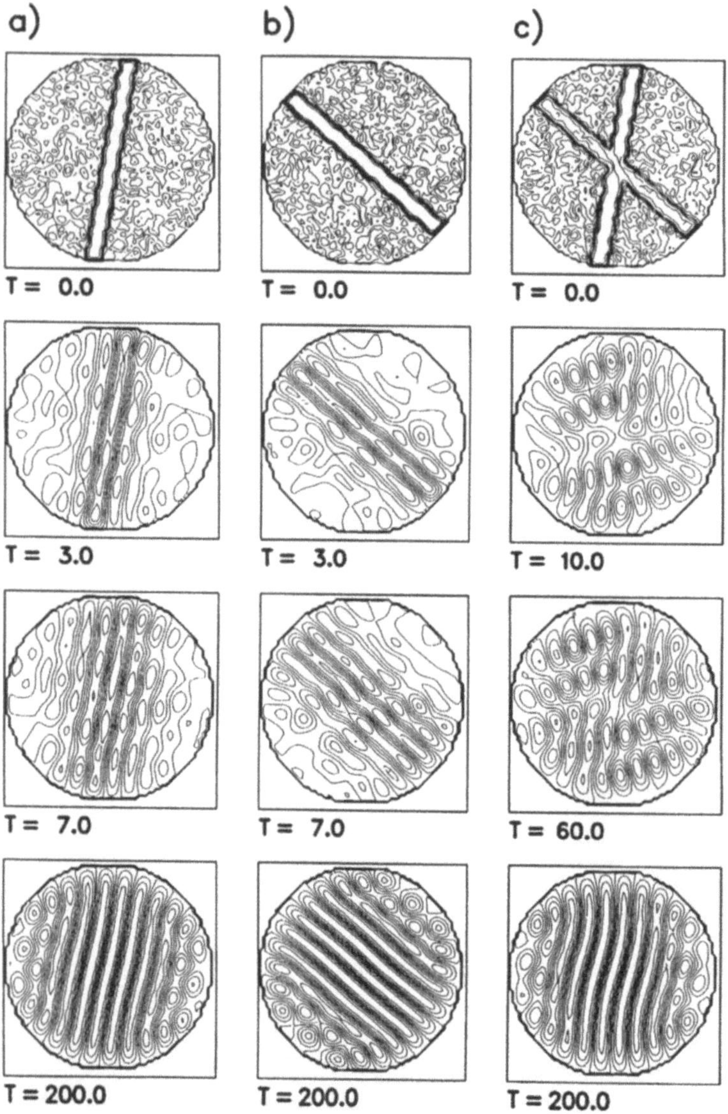

Fig. 7.14. Computer simulation of a fluid heated from below in a circular vessel, where a specific up-welling roll is prescribed initially. Left column: Evolution of one single roll to a full roll pattern by self-organization of the liquid. Middle column: Same as left column, but with a different initial orientation. Right column: Conflict situation in which two up-welling rolls are prescribed initially. The originally somewhat stronger roll wins the competition and thus determines the final pattern

Fig. 7.15. Examples of stored patterns

Fig. 7.16. The computer restores the full pattern out of an initially given part of a pattern

may be formed. First an initial state in form of a single up-welling roll is prescribed. Then according to the computer calculations depicted by Fig. 7.14, left, a complete roll pattern evolves. If we prescribe a different single initial roll, a correspondingly completed roll pattern emerges (Fig. 7.14, middle). Finally, when we prescribe two up-welling rolls (right), one a little bit stronger than the other one, a competition between these two roll patterns sets in and the originally stronger roll pattern wins the competition. In terms of synergetics the following happens: The initially partially ordered state of the fluid gives rise to a set of order parameters. These order parameters compete, whereby the originally largest one wins. It acts on the system by the slaving principle and thus, eventually, forces the whole system to go into its corresponding ordered state. In more abstract terms we may say that a partially ordered system generates its corresponding order parameters that react on the system and force the system into the totally ordered state. Just the same happens in pattern recognition. Once some features are given, they generate the order parameters that forces the total system to complete all the features so that a total pattern is reconstructed. For instance when we prescribe, say, eyes and nose of a person, according to this process the whole face will be reconstructed.

To describe our procedure more explicitly, let us consider a set of faces that have been photographed (Fig. 7.15). In order to process these patterns on a computer, we put a grid over the individual photographs and attribute a grey value to each pixel. The set of all grey values of the stored patterns represents a prototype vector v. Because several prototype vectors are stored, we distinguish them by an index u. Once a test pattern vector q, e. g., eyes and nose, is presented, we have to decide to which stored face this test pattern vector belongs. To this end we construct a dynamics that is described by the following differential equation for the temporal evolution of the test pattern vector

$$dq/dt = \sum_u \lambda_u (v_u \cdot v^u q) - \sum_{uu'} B_{u,u'} (v^u q)^2 (v^{u'} q) v_{u'}. \tag{7.1}$$

According to this equation, the temporal change of the test pattern vector is described by so-called attention parameters λ_u and a learning matrix $v_u \cdot v^u$. The second term on the right-hand side of equation (7.1) serves for the discrimination between patterns and for saturation, i. e., that the test pattern vector values do not exceed a certain value. As it turns out, the temporal evolution of q can be again described by means of a ball rolling down in a potential landscape. As a consequence of this dynamics, in the course of time the originally given test pattern vector is pulled into one of the prototype patterns, which are represented by means of the minima of the potential landscape depicted in Fig. 7.17. Examples of stored prototype patterns and the completion process are given in Fig. 7.15 and Fig. 7.16, respectively. As it turns out, our procedure is very sensitive and allows the computer to distinguish between facial expressions, where examples are provided by Fig. 7.18 and Fig. 7.19. The abstract algorithm described by equation (7.1) can be realized, in particular, by a parallel computer, where each pixel j is connected with a neuron with an activity q_j (Fig. 7.20). Equation (7.1) can then be interpreted in the following way: The neuron with index j receives inputs from other neurons q_k, or from three other neurons with activities q_k, q_ℓ, q_m. The neuron under consideration then sums up over these contributions by means of synaptic strengths λ_{jk} or $\lambda_{jk\ell m}$. This can easily be seen when we rewrite (7.1) in terms of its components:

$$dq_j/dt = \sum_k \lambda_{jk} q_k + \sum_{k\ell m} \lambda_{jk\ell m} q_k q_\ell q_m. \tag{7.2}$$

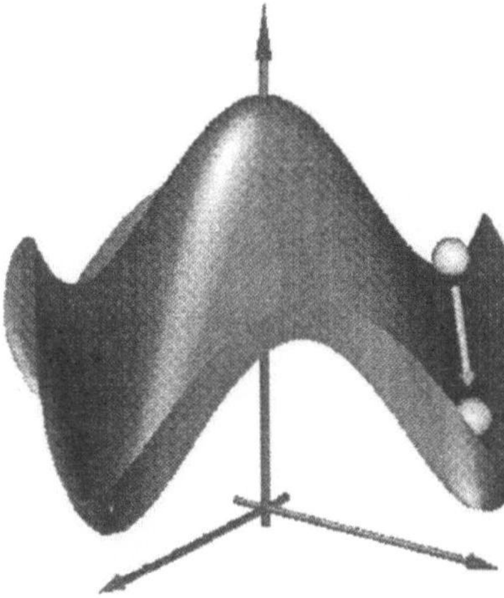

Fig. 7.17. Potential landscape in the case of two features and two prototype patterns. The minima characterize the stored prototype patterns, the ball indicates an initially incomplete pattern

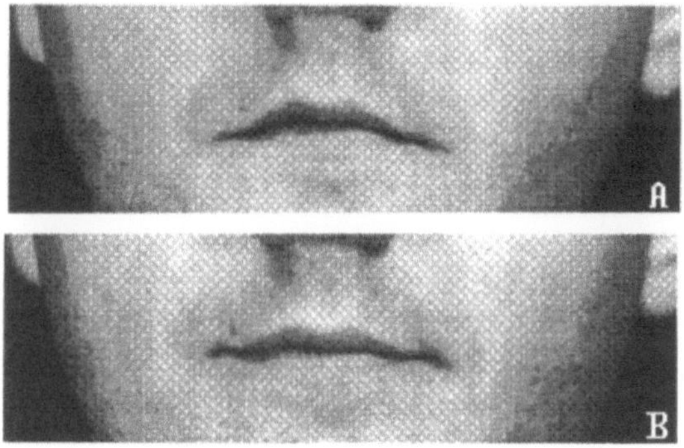

Fig. 7.18. Example for recognition of facial expressions that were successfully distinguished by the synergetic computer

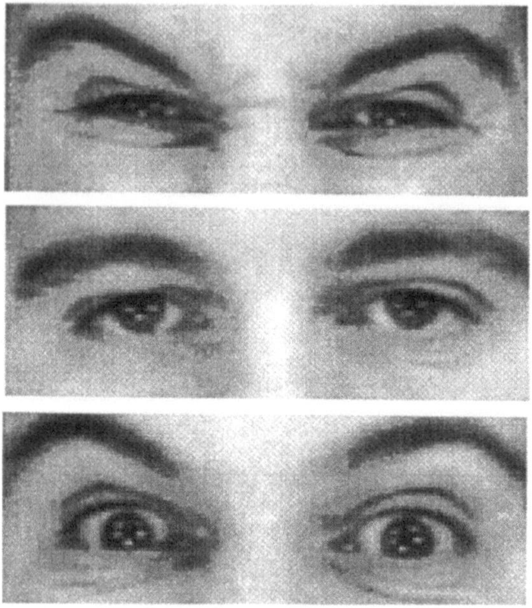

Fig. 7.19. Same as Fig. 7.18, but with respect to eye regions

Fig. 7.20. Realization of the synergetic computer by a parallel network. The neuron with label j and activity q_j receives inputs from the other neurons, sums up and thereupon changes its internal state. These processes go on in parallel

7.13 The Role of Attention. Recognition of Ambivalent Patterns

It is not difficult to develop a preprocessing for the computer so that it recognizes faces that are shifted in their position in space. When we show the computer the scene of Fig. 7.21, it first recognizes the lady in the foreground. Then the attention parameter for the lady is put equal to zero, either from the outside or by the computer. When then the picture is shown again to the computer, it recognizes the man in the rear. Similarly, by setting the corresponding attention parameters equal to zero once an individual face had been recognized, the computer was able to recognize the scene of Fig. 7.22.

Fig. 7.21. Example of the recognition of a scene by the synergetic computer. The lady in the foreground is recognized first, then the corresponding attention parameter is put equal to zero, then the scene is shown again to the computer which now recognizes the man in the rear

These findings lead one to the idea that also in humans our perception is strongly influenced by attention parameters. This view is strongly reinforced by models developed by Ditzinger and the author (1990) in which a saturation of attention parameters is included. If this is taken into account, the oscillations found in the perception of ambiguous figures can easily be explained. An example for an ambiguous figure is presented in Fig. 7.23. The time during which one specific interpretation of an

Fig. 7.22. Same as Fig. 7.21, but containing several faces

Fig. 7.23. Example of an ambivalent figure. If the middle part is perceived as foreground, a vase may be seen, otherwise two faces

ambiguous figure is perceived is called the reversion time. Because in our approach it is connected with attention parameters, one may try to influence the reversion times by means of attention parameters. One possibility that I suggested a couple of years ago was to influence the attention parameters by drugs. In addition one may try to use reversion times to get insight into the emotional state of the test person.

7.14 EEG and MEG Analysis

In the brain both electric and magnetic fields are built up by the cooperation of many neurons. These fields can be measured by electroencephalograms or magneto-encephalograms. More recently it has become possible to analyze some of these measurements in terms of synergetics. Here we wish to represent two recent examples. Using methods of chaos theory, Agnes Babloyantz (1985) showed some years ago that the EEG in the case of epileptic seizures shows chaos of a rather low fractal dimension where the current derivative for one electrode was used. Here I briefly sketch our analysis of EEGs where multiple electrodes were used. The time evolution of a single potential derived from one specific electrode is shown in Fig. 7.24. Figure 7.25 shows the time series correspond-

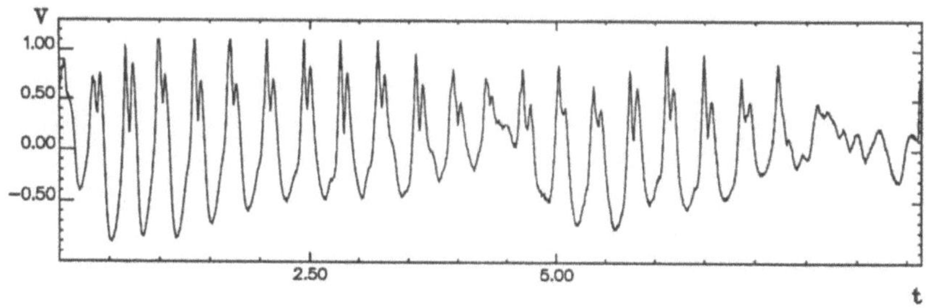

Fig. 7.24. EEG recording: time series of voltage as derived from one electrode

ing to their local positions. From these time series, a spatio-temporal pattern of electric brain activity can be reconstructed according to Fig. 7.26. Each point in time corresponds to a complete figure and its surrounding in that special figure corresponds to a box in Fig. 7.25. If the brain is close to an instability point, according to synergetics, we expect that the spatio-temporal patterns can be represented by a superposition of a few basic patterns. As a detailed analysis reveals, one may indeed represent the patterns of the voltage $V(x,t)$ as a function of space x and time t of Fig. 7.26 as a superposition

$$V(x,t) = \xi_1(t)v_1(x) + \xi_2(t)v_2(x) + \xi_3(t)v_3(x)$$

with three modes $v_j(x)$ only that are shown in Fig. 7.27. The amplitudes are the order parameters. Because there are only three order parameters, their trajectory can be displayed in a three-dimensional space, such as of Fig. 7.28. A detailed analysis by Friedrich and Uhl (1992) has shown

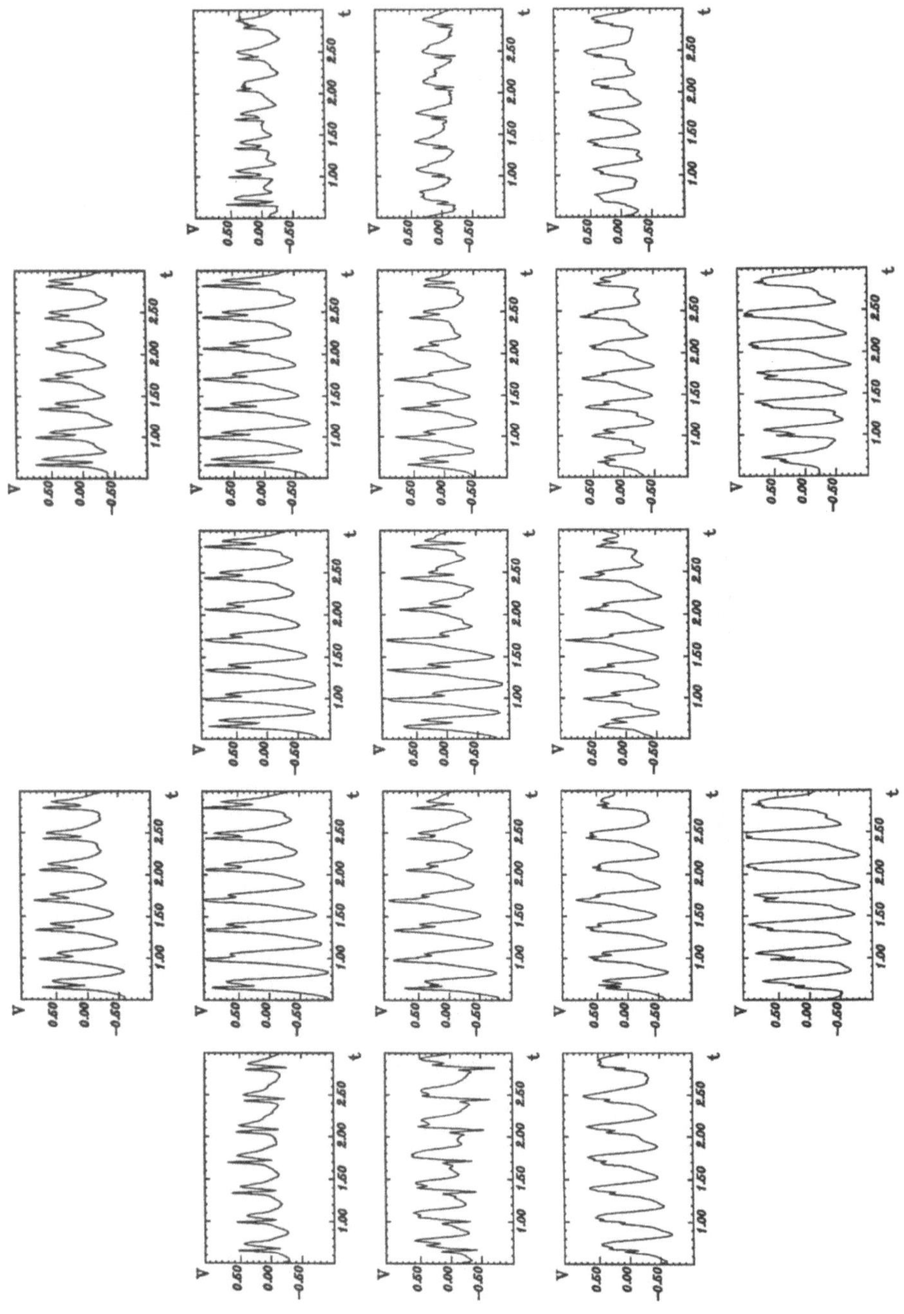

Fig. 7.25. Same as Fig. 7.24, but for all electrodes corresponding to their position on the scalp

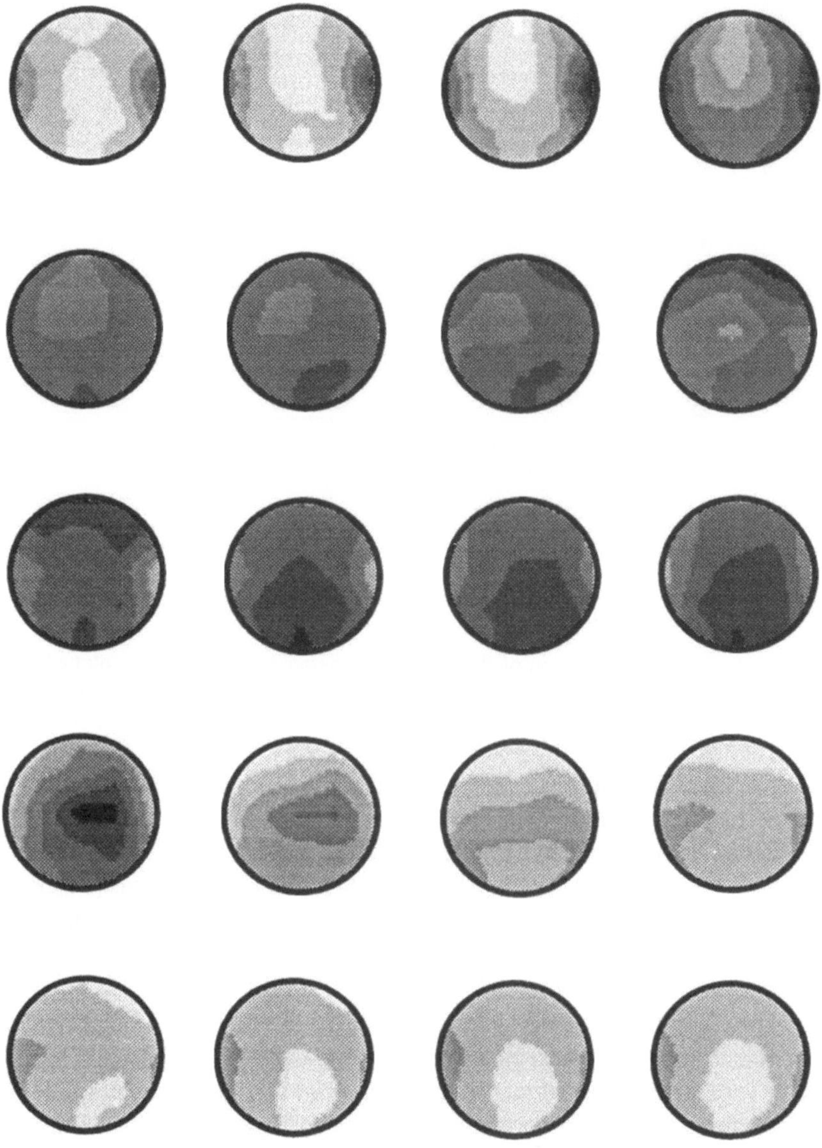

Fig. 7.26. Reconstruction of the spatio-temporal EEG pattern. Each circle corresponds to a spatial pattern on the scalp at a specific time

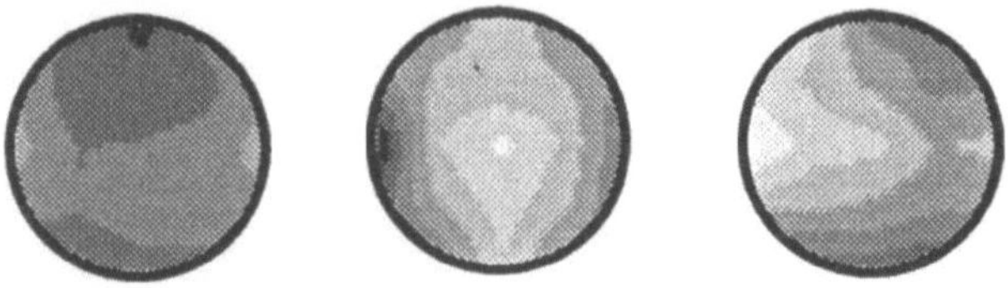

Fig. 7.27. The three basic patterns in the case of *petit mal epilepsy*

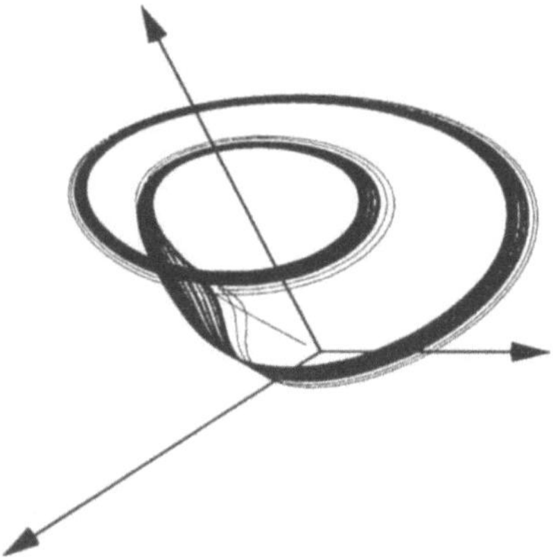

Fig. 7.28. The time evolution of the three order parameters ξ_1, ξ_2, ξ_3 is plotted in a three-dimensional space and shows Shilnikov chaos (Figs. 7.26–28 after Friedrich & Uhl 1992)

that in the case of so-called *petit mal epilepsy*, the trajectory is that of so-called Shilnikov chaos. In this specific case it has been possible to derive a complete dynamical model of *petit mal epilepsy*. We had studied previously α-waves that exist when people are at rest (Fuchs et al. 1987). In this case we showed that the α-waves can be represented by a superposition of only five basic modes that largely correspond to Bessel functions.

This finding shows that interesting brain activities can be represented by few order parameters, or, in other words, in a low-dimensional space. This lends support to our idea that the concept of order parameters and low dimensions plays an important role in the modeling of the brain. While α-waves and *petit mal epilepsy* are so-to-speak free running phenomena, in order to get a more precise insight into brain functions, Kelso and coworkers (1992) performed MEG analysis under given tasks. The typical task was the following: A periodic sound signal is given to a test person who was supposed to react by finger tapping, where finger tapping should

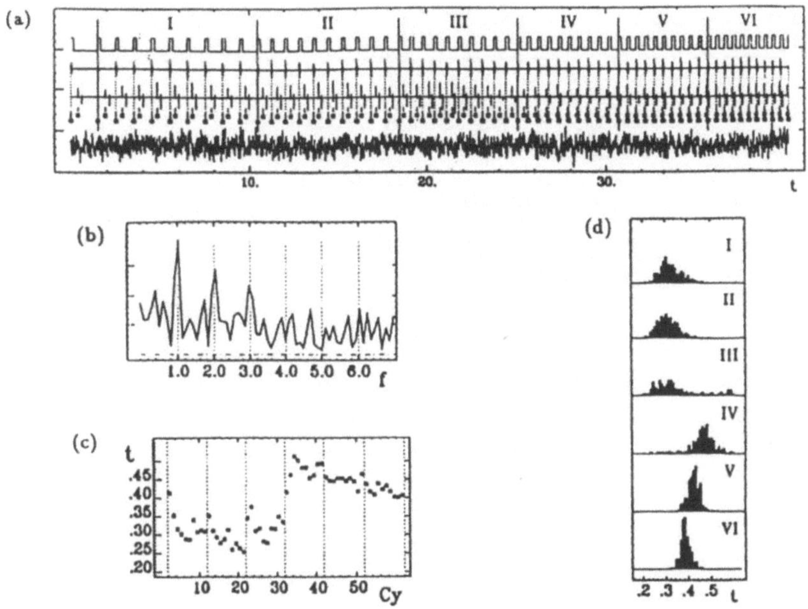

Fig. 7.29a–d. The relation between the stimulus, the response, and the SQUID data from one sensor for a single run. (**a**) 1st row: Plateau number. Points where the frequency of the stimulus changes are marked by solid vertical lines. 2nd row: Stimulus signal, digitized version. 3rd row: Stimulus signal, analog version. 4th row: Response signal recorded from a piezo-electrical switch. The positive peak corresponds to a button press, the negative peak corresponds to a button release. 5th row: Squares showing the phase relation between the stimuli (open squares) and the responses (solid squares). Up to the beginning of plateau IV the response is (approximately) between the stimuli, on plateaus IV–VI the response and the trigger are nearly in phase. 6th row: The SQUID signal from channel 19 recorded during this run. (**b**) Power spectrum calculated from plateau I of the brain signal shown in (**a**). Even though the original time series looks extremely noisy, this spectrum shows clear peaks at the frequency of the stimulus and its higher harmonics. (**c**) Time between the stimulus and the consecutive response plotted for each cycle in (**a**). The vertical dotted lines correspond to a change in stimulus frequency. The transition from a syncopated to synchronized pattern at the beginning of plateau IV can be clearly seen. (**d**) Distribution of the time between stimulus and response on the different plateaus. It can be seen that even as the distribution broadens at plateaus III, the shift shown in (**c**) occurs in almost all runs between plateau III and IV, i. e., at a frequency of 1.75 Hz (after Fuchs, Kelso & Haken 1992)

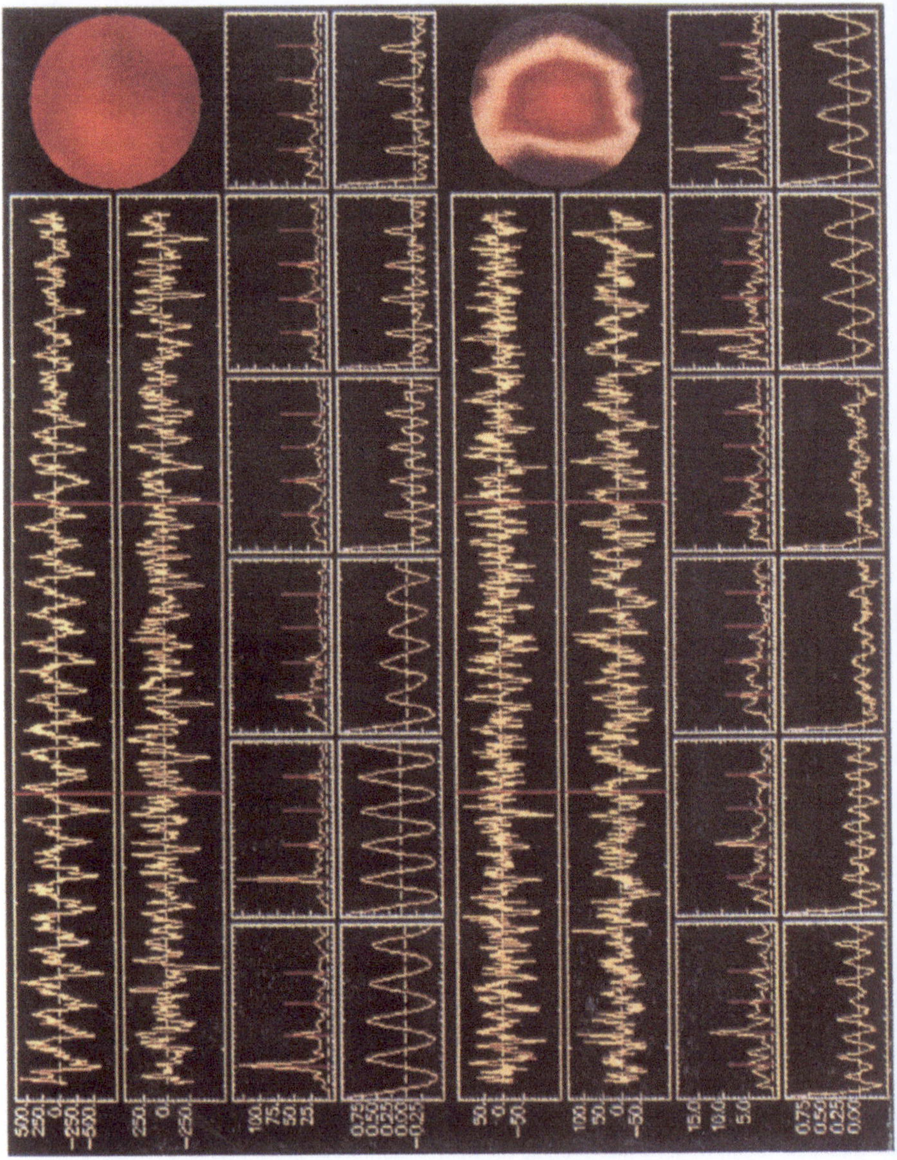

Fig. 7.30a Dynamics of the spatial modes. The mode with $m = 0$: First row: Amplitude of the most dominant mode on plateaus 1–3 (pretransition). Second row: Amplitude on plateaus 4–6 (post-transition). Third row: Power spectra on the different plateaus. Fourth row: Autocorrelation functions for the amplitudes on the different plateaus. Bottom half: The same as above for the sixth mode

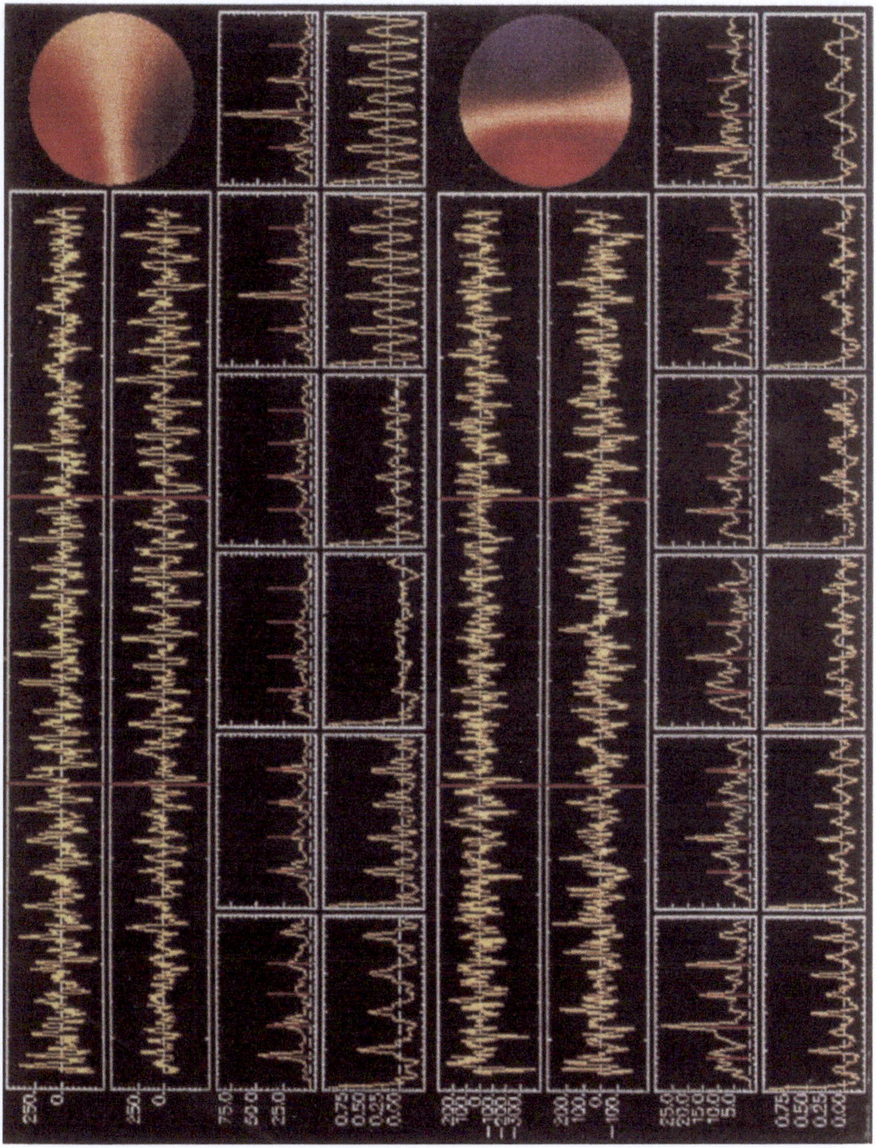

Fig. 7.30b. Dynamics of the spatial modes. The modes with $m = 1$: Upper part: A behavior which is almost the opposite of the behavior of the most dominant mode [Fig. 7.16 a top]. The spectra are diffuse and low in power before the transition, whereas on plateaus 5 and 6 the signal becomes periodic and the spectra show clear peaks. However, in contrast to the most dominant mode, the largest peak is not at the stimulus frequency but at its first harmonic

occur just in the middle between the sound pulses. When the rate of the sound signals was increased, suddenly a change of behavior occurred, namely a transition from syncopating to synchronizing. Again a few fundamental modes were identified and the behavior of the amplitudes studied (Figs. 29, 30) (Fuchs et al. 1992). As it turns out, the process can be modeled by coupled nonlinear differential equations for order parameters (Jirsa et al. forthcoming). These equations represent the phase transition in detail including critical fluctuations.

7.15 Concluding Remarks and Outlook

In the first part of my article I gave a brief sketch of some basic experimental results from brain research. Being composed of a great many of neurons that interact in a complicated manner with each other and producing complicated behavior, the brain is certainly a truly complex system. Thus concepts that enable us to cut a path through this "jungle" are urgently needed. I have tried to demonstrate how methods of synergetics might enable us to deal with this difficult problem. As we could show by a number of explicit examples, the order parameter concept seems to be a useful tool to model the behavior of cooperating neurons at an adequate level. Future research will have to show how this order parameter concept that may easily be extended to hierarchies of order parameters will enable us to treat at least parts of complex brain functioning in an adequate manner.

Pushing our concepts to an extreme, we might say that the thoughts are order parameters and the neurons the enslaved parts. Note that in this way one would say that our thoughts are the outcome of the cooperation of myriads of neurons, but that on the other hand, the action of the neurons is enslaved by our thoughts. It should be mentioned, however, that there are, in my opinion, fundamental limits to the treatment of the brain by these concepts, because qualities, such as perception of color, feelings, etc., are not covered by that approach. There still seems to be a fundamental difference between the action of our brain and, say, that of a machine. Underlying this statement is the idea that all mathematical operations can be realized on the basis of machines. Just think of the concept of a Turing machine.

References

Babloyantz, A. (1985). "Strange Attractors in the Dynamics of Brain Activity" in *Complex Systems – Operational Approaches* ed. by H. Haken, Springer-Verlag, Berlin

Ditzinger, T. & Haken, H. (1989). "Oscillations in the Perception of Ambiguous Patterns", *Biol. Cybern.* 61: 279

Ditzinger, T. & Haken, H. (1990). "The Impact of Fluctuations on the Recognition of Ambiguous Patterns", *Biol. Cybern.* 63: 453–456

Eckhorn, R. & Reitböck, H. J. (1990). "Stimulus-Specific Synchronization in Cat Visual Cortex and its Possible Role in Visual Pattern Recognition" in *Synergetics of Cognition*, ed. by H. Haken, Springer-Verlag, Berlin

Freeman, W. (1991). "The Physiology of Perception", *Scientific American*, pp. 78–85

Friedrich, R. & Uhl, C. (1992). "Synergetic Analysis of Human Electro-encephalograms: Petit-Mal Epilepsy" in *Evolution of Dynamical Structures in Complex Systems*, ed. by R. Friedrich & A. Wunderlin, Springer-Verlag, Berlin

Fuchs, A., Friedrich, R., Haken, H. & Lehmann, D. (1987). "Spatio-Temporal Analysis of Multichannel Alpha-EEG Map Series" in *Computational Systems, Natural and Artificial* ed. by H. Haken, Springer-Verlag, Berlin

Fuchs, A., Kelso, J. A. S. & Haken, H. (1992). "Phase Transitions in the Human Brain: Spatial Mode Dynamics", *Int. Journal of Bifurcation and Chaos* 2(4): 917–939

Gray, C., König, P., Engel, A. & Singer, W. (1990). "Synchronization of Oscillatory Responses in Visual Cortex: A Plausible Mechanism for Scene Segmentation", in *Synergetics of Cognition* ed. by H. Haken, Springer-Verlag, Berlin

Haken, H. (1983). *Synergetics, An Introduction*, 3rd ed., Springer-Verlag, Berlin

Haken, H. (1991). *Synergetic Computers and Cognition*, Springer-Verlag, Berlin

Haken, H. (1993). *Advanced Synergetics. An Introduction*, 3rd printing, Springer-Verlag, Berlin

Hebb, D. O. (1949). *The Organization of Behavior, A Neuropsychological Theory*, John Wiley & Sons, New York

Hopfield, J. J. (1982). "Neural Networks and Physical Systems with Emergent Computational Abilities", *Proc. Natl. Acad. Sci.* 79: 2554–2558

Hubel, D. H. & Wiesel, T. N. (1965). "Receptive Fields and Functional Architecture in Two Nonstriate Visual Areas (18 and 19) of the Cat", *Journal of Neurophysiology* 28: 229–289

Hubel, D. H. & Wiesel, T. N. (1977). "Functional Architecture of Macaque Monkey Visual Cortex", *Proceedings of the Royal Society*, London, Ser. B 198: 1–59

Jirsa, V. K., Friedrich, R., Haken, H. & Kelso, J. A. S. (1994). A Theoretical Model of Phase Transitions in the Human Brain, *Biol. Cybern.* 71: 27–35

Kandel, E. R. (1979). *Behavioral Biology of Aplysia*, Freeman, San Francisco

Kelso, J. A. S., Bressler, S. L., Buchanan, S., DeGuzman, G. C., Ding, M., Fuchs A. & Holroyd, T. (1992). "A Phase Transition in Human Brain and Behavior", *Physics Letters A* 169: 134–144

Kohonen, T. (1987). *Associative Memory and Self-Organization*, 2nd ed., Springer-Verlag, Berlin

McCulloch, W. S. & Pitts, W. H. (1943). "A Logical Calculus of Ideas Immanent in Nervous Activity", *Bulletin of Mathematical Biophysics* 5: 115–133

Rosenblatt, F. (1958a). "The Perceptron: A Probabilistic Model for Information Storage and Organization in the Brain", *Psychological Review* 65(6): 386–408

Rosenblatt, F. (1958b). *The Perceptron: A Theory of Statistical Separability in Cognitive Systems*, Cornell Aeronautical Lab. Inc., Report No. VG-1196-G-1, Ithaca

Stadler, M. & Kruse, P. (1990). "The Self-Organization Perspective in Cognitive Research: Historical Remarks and New Experimental Approaches", in *Synergetics of Cognition* ed. by H. Haken & M. Stadler, Springer-Verlag, Berlin

8. Cognition and the Efficiency of Neural Processes

Hans Liljenström

8.1 Introduction

What Is the Problem?

Living organisms have to survive in a complex and changing environment. This implies, among other things, being able to respond and adapt to environmental events and changes at several time scales. The interaction with the environment, behavior in animals, depends upon the present (dynamical) state of the organism, as well as on previous experiences stored in its molecular and cellular structures. At a long time scale, organisms can adapt to slow environmental changes, by storing information in the genetic material (DNA and RNA molecules) that is carried over from generation to generation. This is referred to as phylogenetic learning. Ontogenetic learning is adaptation at a shorter time scale. It occurs in the non-genetic, primarily neural, structures of the organism, and this information cannot be directly transferred across generations. Figure 8.1 shows roughly how much information can be stored in the genetic material and in the nervous system of various organisms.

At the shortest time scale, the immediate interaction with the environment is partly a result of phylogenetic and ontogenetic learning. Single-cell organisms, such as bacteria and amoebas, have a rather direct interaction between the intracellular processes and the extracellular environment, whereas multicellular organisms have differentiated cells with different functions involved in the interaction. In animals, some cells act as sensory cells, whereas other cells function as motor cells and yet other cells connect and distribute information between the sensory and motor cells. Eventually, larger and larger networks of interconnected nerve cells (neurons) between the sensory "input" and motor "output" expand the realm of behavior of the organism. In man the relationship between sensory cells, brain cells, and motor cells is 10:100,000:1 (Maturana & Varela 1987). It is these masses of interconnected neurons in the brain that make

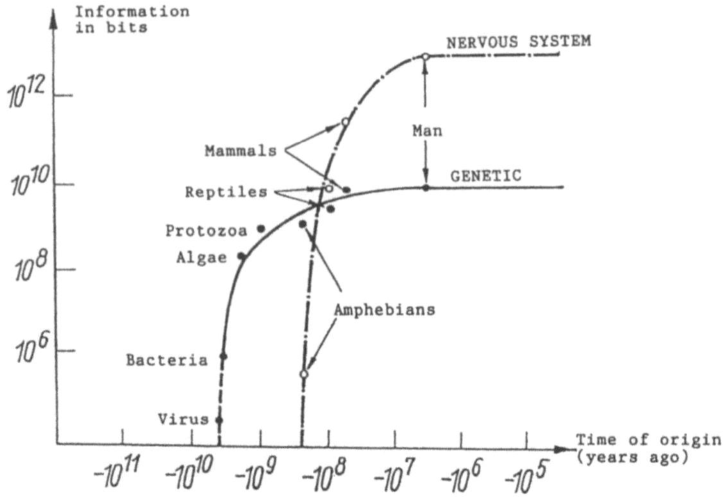

Fig. 8.1. The relative amount of information storage capacity of the genetic material and the nervous system for various species (adopted from Sagan: *The Dragons of Eden*)

our cognitive functions possible; they constitute the material basis for the conscious mind.

In this article, we will discuss the evolution and underlying mechanisms of cognitive functions, and to what extent such functions can be described mathematically and simulated with computer models. Such an attempt requires a clarification of the concepts used. By perception (in mammals), we will mean the ability to form meaningful *internal representations* of sensory information, and associative memory concerns the ability to relate the representation of a current input (pattern) with such previously stored representations. Cognitive functions are sometimes grouped into preattentive and higher cognitive functions, respectively. The latter type of functions are usually considered individual, or "private", in character, implying they are hardly describable objectively. The totality of all cognitive functions could be referred to as the *mind*. Mind can thus be regarded as involving a set of processes, including perception, emotion, memory, thinking, cognition, and reasoning. Some of these may be conscious, while others may not. In either case, typical processes related to mind would require a very high complexity of the nervous system (or any material system keeping such processes going), and cognitive functions could be seen as *emergent* phenomena, arising from the collective neural activity of the brain, or subsystems thereof (e. g., MacKay 1980; Hopfield 1982, 1994; Sperry 1983; Skarda & Freeman 1987; Haken 1991; Heiligenberg 1991). This would imply that these phenomena depend upon, but are not reducible to, events at the single neuron level.

There is no consensus on any definition of either mind or consciousness. Indeed, much of the confusion around this subject depends on the disparate usage of these concepts. For consistency in the description of biological processes, one may adopt the (rather radical) view of Delbrück (1986), and use the terms in a very broad sense. According to this view, consciousness includes all types of "awareness" of external and internal events, from a crude form of perception in the simplest organisms to self-consciousness in Man. The more complex the cellular organization of an organism is the more this organism can be aware, or conscious, of the external and internal world. In this sense, consciousness has evolved together with other properties of life. In his book "Mind from Matter?" Delbrück (1986), regards perception as the basic form of consciousness and he states (p. 43):

> The unity and continuity of life is equally manifest in its psychic aspects. Perception in plants and animals is a familiar phenomenon, but the beginnings of perception are also clearly present in micro-organisms, in which adaptive behavior demonstrates that they can detect and evaluate signals from the environment and respond appropriately.

This is not a dominant view among neuroscientists (or psychologists) today, but it is appealing in the sense that it relates processes in simple organisms with those in higher. It avoids the difficult task of giving a definite dating of the appearance of consciousness in evolution.

While regarding consciousness as a quality that exists to various degrees all along the history of life, we will in the following focus on those cognitive functions that are most prominent for the mammalian brain, in particular (Gestalt) perception, associative memory, and the ability to make plans and predictions of future events. These functions deal with time in a very direct way: the present (perception), the past (associative memory), and the future (prediction). Thinking is an integrative activity that includes a controlled movement of such functions in specific temporal sequences. Experiencing the flow of such activity should be an essential aspect of the conscious mind-process. A major assumption here is that the brain, like other biological systems, has been optimized by evolutionary pressure to efficiently increase the probability of survival in a harsh and rapidly changing environment. This would involve a rapid response to

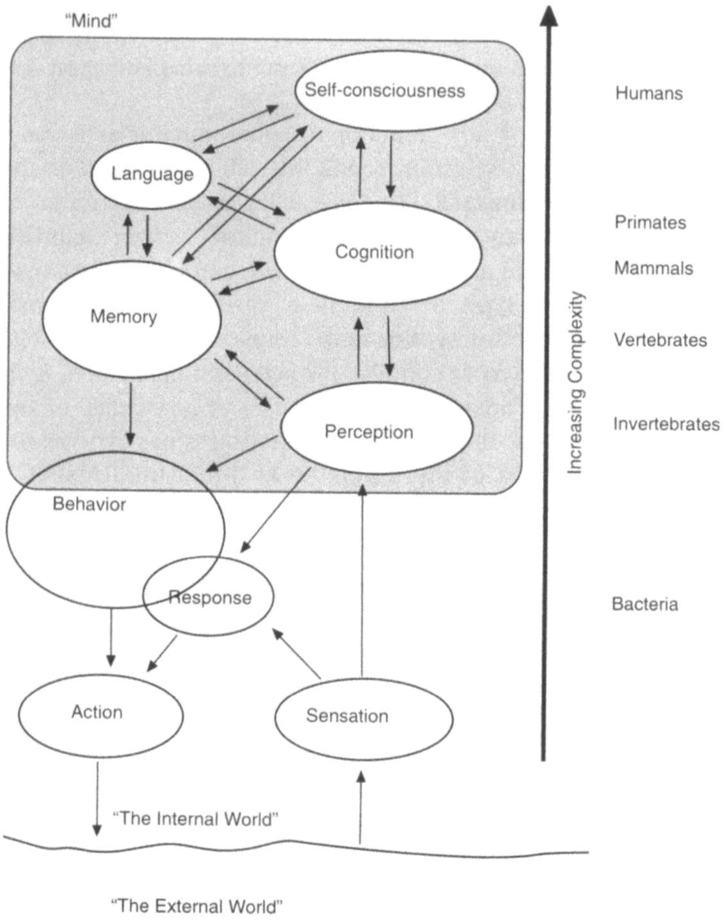

Fig. 8.2. A crude attempt to map various cognitive functions and relate them to each other on an "evolutionary scale" that is at the same time a measure of the degree of complexity

external (and internal) events, i. e., the neural processes of the brain should be optimized with respect to time. What is required in terms of complexity and organization of the nervous system to allow for efficient information processing? What kind of neural dynamics could have evolved for this purpose? Are the same constraints valid for the evolution of "mind-processes" as for other biological processes?

In the following sections we will try to approach these questions, and also show how computer models might be useful in understanding the relationship between neural structure, dynamics, and function. We do not assume that we are able, maybe ever, to model or understand all aspects of conscious mind, such as subjectivity, or understanding itself, but certain mental processes might lend themselves to mathematical modeling and computer simulation. These would primarily be cognitive functions based on neural computations, that could be mimicked in an artificial system. Instead of arguing about which cognitive functions might be based on computable processes, and which might not be, we want to explore how far our computer models are able to carry us in the direction towards an understanding of the brain-mind relation(s). Perception, learning, associative memory, prediction, and even attention are fundamental cognitive functions that people try to describe and simulate with computer models (e. g., Hopfield 1982; Edelmann 1991; Dennett 1991; Haken 1991; Crick 1994). The author's view on how some of these concepts are related to each other and to an evolutionary process is sketched in Fig. 8.2.

As a demonstration, we use a neural network model of the olfactory cortex to investigate the relationship between the neurodynamics and perception/associative memory in that system. The hypothesis is that sensory systems have to provide a fast appropriate response to any particular (in this case, odor) input. Such a response requires a correct perception of the sensory "object" and a decision based on previous experience, i. e., on association. With computer simulations we show that the non-linear dynamics, including oscillations and chaos-like behavior, can give flexibility to the system and speed up the information processing in associative memory tasks. Noise, related to the spontaneous activity of the nerve cells, can also improve system performance. The significance of these findings to brain functions in general, and to mental processes in particular, will be discussed in the last section.

8.2 Evolution of Cognitive Functions

Primary Role of the Nervous System

The nervous system informs the organism about the environment in which it lives and moves, processes the sensory data, relates it to previous experience, and transforms it into appropriate actions or memories. With the introduction of a nervous system, there is a tremendous increase in the way an organism can interact with the environment. In contrast to organisms without a nervous system, those that have such a system can adapt to environmental change and learn by experience at a time scale in the order of the life time of the individual.

Already at early stages in evolution there is perception of the immediate environment. Processing of sensory information can involve discrimination and categorization, and even relatively simple organisms, such as worms, slugs, and insects are able to learn and store information in their nervous systems.

At later stages in evolution, the nervous system is complex enough to allow for "Gestalt" perception, i. e., the ability to "see" an object as a whole. This ability should involve the more or less simultaneous (spatial) *binding* of neural activity in different parts of a cortical area (Crick & Koch 1990; Crick 1994). Although such binding has been discussed primarily with regard to vision, all sensory modalities of higher animals can be said to function with perceptions based on patterns of nerve activities. The ability to form "objects" or percepts of neural activity patterns has apparently increased tremendously during evolution from reptile to mammal. Through such perceptual binding, internal representations of the external world are formed, resulting in an overall "model" of the world. An increased complexity of the evolved nervous system also enables the organism to make better *predictions* about the future environment, based on previous and present experience. The success of these predictions depends to a large extent on the learning and memory capacity of the animal.

At still higher evolutionary levels, reasoning, planning, and abstract thinking come into play. The behavior of the brain (and the rest of the body) becomes less dependent on direct peripheral stimuli, although that will always be of great survival value. Communication and social interaction becomes important for the further development of consciousness (Barlow 1987).

Apparently, the electro-chemical (or other) processes that take place in the intricate neural networks of the brain can lead to various internal states and "mental" processes of an animal. In lower animals the nervous system allows only for "primitive mental" processes, such as perception, learning, recalling, drives, and emotions. To a large extent the animal is unconscious of these internal processes, and it is only aware (conscious, in the Delbrück sense) of the *external* environment. As the nervous system evolves and attains a higher degree of complexity and organization, its processes become more advanced and sophisticated, including cognition, as an act of knowing, thinking, reasoning, believing, and willing. Eventually, (symbolic) language becomes an important and integrated part of the cognitive functions. (Even many of these more advanced activities do not have to be conscious; experiments with so-called split-brain patients show that such activities also can go on unconsciously, at least for the dominant brain half (Sperry 1977).)

In primates, and most articulate in man, the ability to "internalize" the world has come to include the formation of temporal patterns. That is, binding together sequences of neural events that mostly would correspond to some external sequences of events. This *temporal binding* allows us to understand the relation between cause and effect and to experience "the arrow of time". For humans, the internal model of the world also includes other individuals and their minds. Models of other individuals are important for determining the significance of their behavior, and for predicting the "next step" in that behavior. A sense of an "I", or self, can develop when experienced events are put into a sequential context in the inner model of the world. It is obvious that most cognitive functions (but not consciousness in general) depend upon memory, but the "I" needs memories well structured in time, in order to be perceived. Apparently, planning for the future is also an essential component in the making of an "I" (Ingvar 1985). In addition, a certain time span, a few hundred milliseconds, seems to be needed for the "working memory" to create a conscious (subjective) experience (Libet 1985; Crick 1994). We will return to this in the next section.

Although we regard consciousness, in its broadest sense, as a consequence of an evolutionary process, it is possible that some aspects of consciousness, such as subjectivity and the sense of free will, may not have any survival value. They could just be a consequence of the complexity of the human brain. Yet, there are many reasons to believe that a conscious mind actually does have a survival value. For example, it could give an additional motivation to stay alive, to "keep the self intact". By perceiving a self that should live on for many years, there can be better planning both for the individual and for coming generations. It also enables the individual to endure hardships for rewards that may come later in life, hardships that perhaps would be avoided if a future life were not seen. It enables man to behave against "the law of least resistance". Thus, the capabilities to predict future events have been extended to the anticipation and expectation of a continuous individual life. Also, if one has a (good) model of oneself one can infer one's own thinking, feelings, and behavior also in others, thus allowing for a more efficient interaction with other individuals, and better prediction of their behavior. The ultimate pattern in our understanding of the world results in an experience of meaning, a context in which all (or most) events can be ordered and related to each other.

Parts, Levels, and Processes

What part(s) of the brain or what level(s) of neural organization are related to cognition and other functions of the mind? The brain consists of several subsystems and cortical areas, and there are several levels of organization and processing, ranging from molecules to the whole central nervous system. (Processing levels could be described in terms of the synaptic distance from sensory input). Structure and function, as well as type of processing, can vary greatly for different parts of the brain. In "higher" cortical areas, such as the visual cortex or the prefrontal cortex, processing appears to take place in webs of strongly interacting networks. This type of distributed organization may be typical for higher cognitive functions (Churchland & Sejnowski 1988). Visual awareness, for example, has been associated with area V1, but also other cortical areas have been suggested (Crick & Koch 1995). Prefrontal cortex seems involved in temporal processing and is considered to be related to the sense of an "I" (Ingvar 1985). The prefrontal cortex is also the part of the brain that has evolved fastest, and it is uniquely large in man. Yet, even if these brain structures are related to the most human aspects of cognition, neither one of them could be regarded as the "seat of the soul", a concept that seems to be out of place in any respect. It is more likely the interplay of activity in different connected areas that brings about the mind. If we regard mind as a process associated with the overall activity of the brain, no single part of the brain should be expected to be solely responsible for these processes. This notion is also supported by the fact that large portions of the brain can be removed or could be missing from birth, without any apparent disturbance of a person's cognitive capacities.

Yet, even if mind, or any specific cognitive function, cannot be associated with any particular part of the brain, it may be appropriate to ask which level(s) of organization and processing could be involved. What is the relevant spatial scale at which one should look for the material correlates of mental phenomena? Should one go to the microscopic level of sub-cellular processes and structures? Are molecular processes, single ion channel events, single synapses, or single cells important? In that case, could quantum mechanical effects be involved? Do mental processes depend on any particular kind of cells? Do micronetworks, macronetworks, or whole systems form the basis for cognitive processes? Likewise, one could ask which temporal scale would be relevant for the material basis of cognitive functions. (The temporal scale is directly related to the spatial scale, as will be discussed shortly.) At present, there are no definite answers to these questions, but they should be subject to scrutiny.

Temporal aspects are essential for cognitive functions, since cognition must be a continuous process, constantly moving from one "activity" to another. The mind does not (normally) settle for a long period at any one activity, in particular not at rest. It has been envisaged as an "inner eye", or a "searchlight" which moves about inside our heads (Crick 1984). Whatever it is, mind is extremely flexible and adaptive, and for all we know this must be due to a very dynamic brain. In contrast to lower animals with simple behavioral programs, the human brain could not be genetically determined to an extent necessary for its complex interaction with the environment. Instead, in order to enable adaptation to a varied and rapidly changing environment, it should be subject to (structural) changes during development and during the life of the individual. The brain must further be good at predicting changes in the environment, for which purpose some of the cognitive functions can be assumed to have evolved. For this and other reasons it is necessary that rapid, non-permanent changes also take place at a time scale much shorter than the life-span of the individual. We will return to this in the next section.

From an evolutionary point of view, it seems that cognitive functions would appear as a result of the increased complexity of the nervous system. Even if one adopts Delbrück's view that there is some degree of awareness (perception) in unicellular organisms, the kind of "higher" cognitive functions we are discussing here should depend on the co-operative activity of a large number of interconnected neurons. Processes at subcellular levels have hardly evolved radically since the first appearance of multicellular organisms. In this respect it seems unlikely that cognitive functions should crucially depend on microscopic events and structures at such levels where quantum effects might play a role. It is also reasonable to assume that the relevant dynamics of the neural structures have evolved to cope with events and changes in the environment of the organism, i. e., at a "macroscopic" time scale. Finally, since such functions as perception and associative memory supposedly involve neural activity of rather large cortical areas, corresponding to millions of neurons, it can be concluded that one should look for the material basis of cognitive functions at the level of neural networks. Networks must be crucial if mental processes are associated with patterns of neural activity. This assumption will be guiding throughout the remainder of this chapter.

8.3 Efficiency of Neural Processes

During the course of evolution life processes have become efficient in making different organisms survive in a complex interplay with each other and with a changing environment. Efficiency in this context, and in a very broad sense, could mean the ratio of the "output" quantity, e. g., the production of viable offsprings, to the "input" investment of energy and mass per unit time. The demand for efficiency should also be decisive for the brain, but the question is, in what way? How should we measure or define efficiency in neural processes?

Information Processing

Presumably, the nervous system has evolved to process and store "neural information" in an efficient way. This could mean that it has been optimized for a maximum processing rate, or a maximum information storage capacity (or both). Although one often discusses how much information can be stored in the brain, it seems likely that the nervous system has evolved primarily for fast information processing, from sensation to action. A high information storage capacity is supposedly of a secondary, and phylogenetically younger origin, evolved for more advanced behavior in a more complex interaction with the environment.

For a simple nervous system, such as that of an insect, the capacity for storing new information is low. Instead, most of its wiring is determined genetically, which is possible for such simple but life sustaining actions as feeding, mating etc. The action control program can be specified by a small number of instructions. For example, a single specific molecule, such as a sex pheromone, can trigger the flight instructions and the mating behavior of an insect. In the more complex nervous system of, for example, a mammal, there is no way to genetically encode all the possible behaviors in the natural environment of such an animal. Instead, there must be a great capacity for learning and adaptation during the lifetime of an individual. An increase in the number of cells and synapses, and hence in the number of processing steps, will presumably make the system slower, but will also allow for a richer and more robust behavior.

The word "information" should be used with caution. Often, when discussing the brain's capacity for signal/information processing, one treats the environmental input as if there were many signals (implying information) with noise superimposed. Then, one discusses the capability for increasing the signal-to-noise ratio. However, the information of the "environmental signal" only becomes apparent, in some sense is *created*, in the brain. Much of the sensory input is "nonsense", until it reaches higher cortical areas where it attains meaning. What the brain does is to make meaning out of sensory input (cf. Freeman 1991; Maturana & Varela 1992). This process involves a lot of filtering, clustering, separation, association, etc. While information alone can never produce mind, the organized amount of information processed per unit time seems to be important. A high flow rate of information may not be sufficient for mind, but it is very likely a prerequisite for it. Also the (dynamical) state of the brain is of great importance for how the information is processed, and what the result of this processing will be.

Optimization

In biological processes in general, there is an interdependence, or trade-off, between energy, accuracy, and time (rate), and different processes seem to be optimized with respect to either one of these. This is perhaps most clearly seen in cellular processes, such as protein synthesis (Liljenström 1987; Blomberg & Liljenström 1988). Depending on whether the main "objective" of the process is to ensure a "production" rate that is as high as possible (e. g., as in bacterial cell growth), or if high accuracy is needed (as in DNA replication), or alternatively, if the energy consumption needs to be minimized (as in starvation), the processes should be optimized with respect to time, accuracy, or energy, respectively. This optimization also determines the efficiency of the system.

If, for example, energy is abundant or the energy consumption is fixed, time and accuracy could be linked in such a way that an increase in accuracy would require an increase in processing time. Time is presumably the most important factor in many real-world situations, i. e., a rapid response is favored at the expense of lower accuracy (a too-low accuracy could of course be fatal). In other situations, accuracy could be more important. It is conceivable that some controlling mechanisms have evolved to shift the efficiency constraints of a process depending on the environmental constraints.

Also when it concerns other biological problems, such as to find an appropriate balance between stability and flexibility (sensitivity) in, e. g., genetic or neural processes, optimization principles seem to be at work to yield efficiency. How much of the information should be "hard-wired" and how much should be learnt by experience? What is an optimal balance between evolutionary (phylogenetic) learning and ontogenetic learning in any given species and environment? What, if any, would be the constraints that determine the optimization principles at work? Under what evolutionary pressure is the nervous system? We will briefly consider some aspects on energy, accuracy, and time related to brain processes, and suggest that, in general, it should be important to minimize time also here.

Energy

In all physical systems energy is a fundamental quantity. In biological systems, energy is invested in large more or less stable molecules, as well as in smaller elements (molecules and ions) used for driving the processes. For example, in protein synthesis a large amount of energy is invested in nucleic acids (RNA) and in polypeptides, whereas nucleotide triphosphates (e. g., ATP) are used in a more direct energy transfer for driving the process. In the nervous system, the relevant energy consumption, apart from the pure maintenance of the cellular processes, is in making new and stronger connections between neurons and in the ion pumping used for action potentials and transmittor release in synapses, both processes essential for information transfer.

Glucose is virtually the sole energy resource for the human brain (except during prolonged starvation). The brain requires a continuous supply of glucose and consumes about 120 g daily, corresponding to an energy input of about 420 kcal (or 1760 kJ), which gives a power of about 20 W. The brain accounts for some 60% of the utilization of glucose and about 25% of the total energy used by the whole body in the resting state. The blood glucose level varies during the day, but stays above 50 mg/100 ml even during starvation, since the brain cannot tolerate appreciably lower glucose levels for even short periods. In fact, the first priority of metabolism (in starvation) is to provide sufficient glucose to the brain (Stryer 1981). Under certain circumstances, energy supply might be a limiting factor for the brain functions, but under normal conditions energy should be abundant. Levy and Baxter (1995) have discussed energy efficiency of neural coding, but here we will only be concerned with cases where energy is not limiting.

Accuracy

Accuracy, fidelity, or quality, are concepts that have a meaning only to biological or man-made systems. A high accuracy is essential for the life processes, from DNA replication to neural information processing. Yet, the accuracy of these processes should not be infinite, even if that could be achieved. In the replication of DNA, errors are crucial to allow for evolution and inheritable adaptation to environmental changes. Likewise, it is conceivable that errors are important in the neural information processing. Such errors could be due to the spontaneous firing of action potentials, or to synaptic noise (cf. Århem & Lindahl 1997).

Even though noise and errors are inevitable, and under normal conditions may be harmless, accuracy must be kept at an adequate level to ensure a functional information transfer. A crucial question is how accurate the neural information transfer needs to be to support an appropriate behavior of the animal. It is reasonable to believe that the system has evolved to function optimally at some range of error rates. However, it is much harder to define accuracy, or the error rate of a neural information transfer than, e. g., of DNA replication or protein translation, where the code is physically present as a large polynucleotide molecule. There is no similar code in a physical form in the neural processes. One attempt to measure the error rate in this case is to register the occurrences of events, such as the firing (or not firing) of an action potential, where such an event "normally" should not (or should) occur if the membrane potential is well below (or above) the threshold for firing. This kind of events could also include the release and reception of transmittor substances in the synapses. The problem is again the stochastic uncertainty which makes a clear definition of "normally" impossible. At a more "global" scale, an error can be related to an action in response to any particular stimulus, for instance "eat or flee" at the sight of an animal of a different species. This kind of mistake can naturally be fatal for the individual. Hence, one can expect a higher probability for "over-reaction", i. e., for responding too often rather than failing to respond. Naturally, in real world situations where the sensory input is very complex and noisy and where thousands or millions of neurons are involved in any activity, it is difficult to talk about the accuracy of a single neuron or neural process. It is the overall effect that counts, and there may be many reasons why the system does or does not respond the way it "should".

The role of noise, or random events, in biological systems was perhaps first brought to attention by Schrödinger (1947). In his book *What is Life?* Schrödinger asks fundamental questions regarding the stability and sensitivity of our body in general, and of the brain and sensory organs in particular. He argues that our sense organs (and the brain itself) would be useless if they were too sensitive and reacted to single atomic motions. Schrödinger points at a general statistical law, the so-called $\sqrt{n}$ law, as a measure of the degree of inaccuracy to be expected in any physical law. The relative error of such a law (of physics or chemistry) is of the order of $1/\sqrt{n}$ (i. e., $n^{-1/2}$), where n is the number of molecules that co-operate to bring about that law. Hence, if $n = 1$ million, the relative error in any measurement would be 1/1000, or 0.1%. From this one can see, as Schrödinger argues, "that an organism must have a comparatively gross structure in order to enjoy the benefit of fairly accurate laws, both for its internal life and for its interplay with the external world. For otherwise the number of co-operating particles would be too small, the 'law' too inaccurate".

A similar line of arguments could perhaps be applied also to single action potentials (APs) or other events dependent on single cell activities. If we could determine the number of APs (or active neurons) involved in, say, the perception of an object, the $\sqrt{n}$ law would give the inaccuracy of that particular brain process. Conversely, if a certain degree of accuracy is needed for any particular (cognitive) process, the same law would give the approximate number of events, or "particles", necessary to perform the process.

Since there is continuous spontaneous activity, noise, in the brain, it should not normally be sensitive to single APs, as discussed in general terms by Schrödinger. The activity of single cells appears to be largely unpredictable and noisy, but the mass of cells co-operates to produce a coherent pattern. It is the mass action of thousands of cells that make the orderly dynamics necessary for cognitive functions (cf. Freeman 1991). However, there may be situations where spontaneous neuronal events, such as the opening of a single ion channel, can be amplified, resulting in an AP, that in turn can result in a cascade of neural activity. (see Århem & Lindahl 1997) It is worth noting that Delbrück and others, who argue against the biological relevance of quantum events, regard such amplifications of microscopic events or fluctuations as possible exceptions (Delbrück 1986). Another "macroscopic" effect could be, as shown by computer simulations discussed later on, that noise can result in a switching between different dynamical (attractor) states.

Time

Time has also a very special meaning to biological processes, not only because of the irreversibility of life processes. Environmental changes occur at several time scales relevant to life, but it is impossible to make any exact separation of these scales matching the different life processes. One attempt is to crudely relate them to the average life span of an individual organism. In particular, this can be applied to the animal brain.

The "dynamics" of the brain can be considered at three different time scales: a large one based on genetic changes (evolution), an intermediate one based on synaptic changes (learning), and a small one based on neuronal activity changes, directly related to cognitive functions. The latter should be on the order of a few hundred milliseconds, or less. Of course, there are molecular and atomic processes in the cells which act on a much smaller time scale (nanoseconds–milliseconds), but these are likely to be of only indirect relevance to the "macroscopic" events and processes of the organism.

For fast changes (compared to the individual lifetime) it is necessary to have a low energy cost per event, implying small "molecules". This, in turn, may result in a low accuracy per event, and hence a large number of events, or "molecules", are needed to achieve an acceptable accuracy. In general, a higher rate is achieved at the cost of a lower accuracy and/or a higher total energy input, but as discussed above, the accuracy cannot be allowed to become too low. There is also an upper limit to the energy input. Hence, there has to be a "trade-off" between time, energy, and accuracy. As already discussed, in many natural situations it is important to minimize processing time (maximize the rate). It seems reasonable to believe that a major constraint on the nervous system and its dynamics is to maximize reaction rate upon any given sensory input. This assumption underlies the discussion and the modeling efforts presented in this article. Table 8.1 gives an approximate separation of time scales for life processes, and how these processes are related to energy consumption and accuracy in information transfer.

If the relevant information transfer concerns events and processes at a sub-cellular level, the appropriate time scale could be very small, on the order of nanoseconds or less. This is the time scale of conformational

Table 8.1. An approximate separation of time scales for some life processes, as related to the life span (l.s.) of higher organisms, such as mammals. The relevant "particles", the energy and the accuracy involved in the processes are also listed.

Adaptation to environmental change				
Process	Time scale	"Particles"	Energy	Acc
Evolution	$10^4 - 10^7$ yrs $\gg$ l.s.	DNA, RNA	high	high
Learning	days $- 10^2$ yrs $\approx$ l.s.	proteins, lipids	medium	medium
Neurodynamics	msec – min $\ll$ l.s.	ions, peptides	low	low

changes in proteins, ion channel openings, etc. On the other hand, if the relevant information transfer is not within but between cells, the upper limit for the rate of information transfer is determined by such parameters as interspike interval (the inverse of firing rate), membrane time constant, and synaptic and axonal conduction delays. The values of these parameters vary for different types of cells and fibers and signal traveling distance, but are all in the range of one to tens of milliseconds. The shortest time for information transfer, from sensation to action, would be found in simple nervous systems, such as in insects. It is known that, e. g., flies can react to a single cell response and change their direction of flight within a few milliseconds, as a reaction to a single or a few action potentials (Downer 1988). Certain ants are supposed to catch their preys within a single millisecond upon detection. In larger networks, such as in the mammalian brain, where normally millions of cells are involved in any type of activity and where there are several synaptic steps between sensory system and motor control, the shortest time is much larger, typically on the order of a hundred milliseconds or more. Some specific pathways, like those involved in simple instinctive sensation-reaction schemes, may be very fast. In those cases, however, the process would probably not be considered as a cognitive process.

Several hundred milliseconds seem to be needed in order to build up a "conscious experience" of a neural event (Libet 1985). During this time, stored memories may be activated to modify the incoming impulse, or to put it in a context that is necessary for the existence of a conscious mind, or an "I". However, the organism (animal) may very well function without this consciousness. Some reactions that are not only pure reflexes may still require less time than half a second. The brain seems to be able to act in a relevant and adequate manner, even without "consciousness". For example, when a cheetah is hunting an antelope, it has to make decisions within a few tens of milliseconds. If it is running at a speed of 100 km/h, this corresponds to approximately one meter in 35 milliseconds. A turn or a jump has to be done at the right moment, based on decisions made within this time interval. Similarly for a skier in a downhill race. It is not just a question of reflexes, involving only the spinal cord, these decisions must involve some sort of brain activity. (See Dennett, 1991, for further aspects on time and consciousness.)

8.4 Neurodynamics

Close to Instability, "At the Edge of Chaos"

If the nervous system, in particular the brain, has been optimized during evolution to deal with complex and rapid environmental changes at time scales shorter than or comparable to the life span of the individual (including interaction with other individuals), this should be reflected by a rich "inner" dynamics. Let us look at the kind of neurodynamics that could account for the efficiency of the organism and increase its chance of survival, both at a very basic level and at the level of higher cognitive functions.

Whereas biological systems are complex and driven far away from equilibrium, most systems described mathematically in e. g. physics and chemistry are relatively simple and in equilibrium. In contrast to systems in equilibrium, where "nothing interesting happens" – the future looks the same as the present – complex systems far from equilibrium can exhibit a very rich dynamics. The future is here unpredictable, but with information about present and past states, future states can in some cases be predicted to a satisfactory degree. This is done with the aid of the dynamics of the brain, which also determines the complex behavior of the organism. The activity of the brain can partly be studied with experimental techniques, such as electroencephalography (EEG) and positron emission tomography (PET). They reveal a highly complex dynamics, with oscillations in certain frequency bands interspersed with aperiodic, chaotic-like behavior.

The rich dynamics of the brain can be well exemplified by the olfactory system (primarily bulb and cortex), which has been extensively studied by Freeman and co-workers (1975, 1987). This system processes odor information, determining the quality and quantity of odor objects in a fluctuating environment. An essential feature in its dynamics is spatio-temporal patterns of activity, which do not seem to depend critically on the detailed functioning of individual neurons. Self-organization of patterns appear at the collective level of a very large number of neurons. Oscillations occur at various frequencies, in particular around 5 Hz (theta rhythm) and 40 Hz (gamma rhythm). There are also waves of activity moving across the surface of the olfactory cortex. In EEG studies of bulb and cortex Freeman (1991) has also found evidence for chaos, or at least aperiodic behavior different from noise and with some degree of order.

What is the significance of this dynamics? How is it related to computation and information processing in the brain, and in particular, to cognitive functions? As pointed out before, cognition requires a rich dynamics, but the relationship is far from clear. In some way, the dynamics of the brain, as "the inner world", seems to be able to reflect the dynamics of the environment, "the external world". Although oscillations and "chaos" could be mere epiphenomena, as an inevitable consequence of the neural circuitry and without any specific biological role, these phenomena have been suggested to have functional significance.

A possible role for the gamma frequency oscillations could be a means to compare incoming stimulus with previously stored patterns (see, e. g., Wilson & Bower 1989). The slow theta rhythm could serve to hierarchically trace down the olfactory information in a feedback loop between the olfactory bulb and cortex (Ambros-Ingerson et al. 1990). Yet another possibility is that oscillatory bursts at this low frequency, approximately 5 Hz, would allow the system to relax to an attractor (limit cycle) within a 200 ms cycle, without disturbance of successive sensory information during the interburst intervals. The oscillations may also have a more direct role in learning and associative memory functions. It has been shown, for example, that theta rhythm oscillations are optimal for induction of long term potentiation (LTP). The findings of strongly correlated stimulus-invoked oscillations at approximately 40 Hz in the cat visual cortex (Gray et al. 1989; Eckhorn et al. 1989) have led to a theory for visual awareness (as part of consciousness) that would solve the so called binding problem (Crick & Koch 1990; Crick 1994). Separate cell assemblies responding to different aspects of an object could be "labeled" by frequency and/or phase to form the perception of one single object. This would suggest a mechanism for the "searchlight hypothesis" of mind, previously proposed by Francis Crick (1984). Yet, the simplest and perhaps most direct role of the cortical oscillations could be to enhance weak signals and speed up information processing.

It is more uncertain what role chaos, if it exists in the brain, could have for neural computation, where convergence, not divergence, should be desired. However, the chaotic-like behavior found in certain brain structures could serve a purpose if it is transient or intermittent between stable (dynamical) states. It seems plausible that the chaos-like behavior of a neural system, such as the olfactory bulb and cortex, could yield flexibility to the information processing, by allowing the system to rapidly shift from one complex activity pattern to another in response to a small input. At a higher level it could be responsible for the brain's capacity to generate novel activity patterns, corresponding to its internal self-generated ("creative") thought processes (Freeman 1991). Several other roles for chaos in neural systems have been suggested (see Tsuda 1991), but again, maybe the most important one is to increase flexibility and reduce processing time.

Neuromodulatory Control

In order for the system to make use of the various dynamical states there has to be some kind of regulatory or control mechanisms. Many factors influence the dynamical state of brain structures, for example the excitability of neurons and the synaptic strengths in the connections between them. A number of chemical agents act on these neural properties. Such agents, e. g., acetylcholine (ACh) and serotonin (5-HT), can change the excitability of a large number of neurons simultaneously, or the synaptic transmission between them. They can thus be regarded as control parameters. The concentration of these "neuromodulators" is directly related to the arousal or motivation (or mood) of the individual, and can have profound effects on the neural dynamics and on memory functions. For example, ACh increases the oscillatory activity in the olfactory cortex and in brain slices of hippocampus. In addition, low levels of ACh have been found to accompany memory impairment in Alzheimer's syndrome (see refs. in Liljenström & Hasselmo 1995).

The central idea that has been discussed here is that the complex dynamics of the brain is a consequence of an evolutionary pressure to cope with a complex and rapidly changing environment. The internal processes of the brain presumably reflect the external processes with which they interact. In the following, we will show how the regulated neurodynamics described above can be simulated with computer models and demonstrate how it can give an efficient information processing.

8.5 Computer Models

Models of Cognitive Functions

Is it possible to test the ideas presented here using computer models and simulations? Could computer models help us to understand the relationship between structure, dynamics, and function? Is it at all possible to use such methods as tools for understanding the mechanisms behind cognitive functions? Indeed, many authors believe this could be a fruitful, although not conclusive path. Not only computer scientists working with brain models, but the view is also shared by a great number of neuroscientists, philosophers, and others (Edelman 1988; Haberly & Bower 1989; Crick & Koch 1990; Dennett 1991; Churchland & Sejnowski 1992). Others are more sceptical (Penrose 1989; Roland 1993) and stress that any model must be "ontologically committed", meaning that all variables and parameters introduced in a model should have a counterpart in the real system (Roland 1993). This is an important point that should be taken seriously. However, with an appropriate computer model we can make simulation "experiments" that would be inconceivable, or would take an unrealistic long time to carry out with the real system. For example, if a model parameter can be related to the concentration of a certain transmittor substance or neuromodulator, its effects on the dynamics and functions of the system can be investigated in detail using computer simulations. In the following we shall look at some of the questions the modeler has to address, and we shall also give an example of a particular type of neural network model used for some of the basic problems involved. For a more thorough discussion on the possibilities and limitations of computer models in neuroscience, see Lansner and Liljenström (1994).

When making a computer model of a biological system like the brain, or some part of it, one has to find an appropriate balance between realism and abstraction. The amount of available data of the structure and function of the system is far too large to be included in any one model or simulation. Even when the aim is to mimic a system as closely as possible, there is an inevitable need for simplifications. A model whose main purpose is to capture the crucial features of a complex system should be made even simpler. The challenge is to make the proper simplifications, extracting those features of the system that are essential for the particular aspects one intends to model. If a model is simple enough and only has a few parameters, analytical solutions may be found. However, in most cases where some realism is sought, the models become far too complex, incorporating thousands of equations and many more parameters. Then, only numerical solutions are possible. Simulations of large and detailed mathematical models are time-consuming, but with the computational power of today's computers, this is no longer a serious bottleneck in computational neuroscience.

What structural properties are responsible for the dynamic behavior described in the previous section? Which type of networks or systems possess the properties necessary for cognitive functions? What kind of algorithms are involved? Computer models may aid in answering these questions, in particular when exploring possible (and impossible) relations between structure and function, and determining system properties/parameters relevant for any specific phenomenon.

If cognitive functions primarily depend upon network properties, it may be adequate to let the network nodes represent large populations of neurons. In this case, one could describe mathematically the average output activity of this large set of cells. This output can be described in terms of pulse density, or mean firing rate, rather than in terms of action potentials, resulting in a continuous non-linear input-output relation, usually described by some sigmoid curve. Details at the level of single cells or synapses (or even lower levels) may still play a role, but it is questionable if the system at a large scale is accurate or sensitive enough to depend crucially on these "microscopic" objects and events.

Besides the characteristics of the network nodes, the properties of a neural network is determined by the network connectivity, and how the connections between cells (nodes) can be modified. Donald Hebb (1949) was a pioneer on this subject, and his ideas are still valid. One of his key ideas was that repeated and near simultaneous activity in pre- and postsynaptic ("sending" and "receiving") neurons results in a strengthening of excitatory synapses between these neurons. This is what today is referred to as Hebbian synapses, and their existence is now experimentally established and generally accepted (Malenka & Nicoll 1993). By the action of such synapses, neurons activated by a certain stimulus will be tied together by strengthened mutual excitation and will form what Hebb called "cell assemblies". Each such assembly will serve as an internal representation of the corresponding object, or concept, and its existence greatly influences the system dynamics. Further, temporal sequences of active cell assemblies have been proposed as the physiological correlate of thought processes. Hebb's theory is still in its major aspects compatible with experimental findings on cortical architecture and function and represents a fruitful starting point for theories of cortical function (Palm 1990; Singer 1990).

There is a strong correspondence between Hebb's original model and so-called attractor neural network models, the most well-known of which is the Hopfield net (Hopfield 1982, 1984; Amit et al. 1991). Such a network has the same structure as proposed by Hebb, and if it is trained using a correlation based Hebbian learning rule it will produce an associative memory with similar characteristics as those suggested by him. In a trained attractor network, the connections between units representing parts (features) of a memorized pattern are reinforced. Those units that never or rarely occur in the same internal object representation develop mutual inhibitory connections. The resulting network connectivity determines the structure of the network, and thus its dynamics and pattern reconstruction capability.

Olfactory Cortex as a Model System

Most connectionist models so far are based on the convergence to steady state "point attractors" (as in the Hopfield net), which may not be appropriate for any real neural system with a rich dynamics. The different (functional) "states" of the brain would then make up a multidimensional "energy landscape" (Hopfield 1982). In the simplest case, this landscape is supposed to be more or less fixed, with valleys and ridges statistically determined by the network connections, and where the valleys correspond to point attractor memory states. However, a more realistic picture would be that of a roaring sea, which constantly is changing, and where the memory states rather would correspond to the rolling waves.

A more complex dynamics with oscillations and chaos-like behavior is found for models with a more realistic architecture, for example in models of the olfactory system (Freeman 1975; Skarda & Freeman 1987; Li & Hopfield 1989; Wilson & Bower 1989; Liljenström 1991; Erdi et al. 1992). Compared to, for example, the visual system, the olfactory system (primarily the bulb and the olfactory, or piriform, cortex) is more primitive and much simpler. The visual cortex has six layers and is about ten synapses away from the retinal receptors. The olfactory cortex, on the other hand, consists of only three layers and is only two synapses away from the odor sensory receptors. The connectivity patterns of this cortex is also of a more distributed character, with extensive short and long range connections within a layer. It therefore resembles more the abstract models of associative memory than does the visual cortex. Both from its structure and its rich and well-studied neurodynamics (see previous section), the olfactory cortex can be regarded as a suitable model system for a connectionist approach (Haberly & Bower 1989).

A Neural Network Model of the Olfactory Cortex

In order to investigate whether the neurodynamics of cortical structures has any significance for perception and associative memory a computer model of the olfactory cortex was developed (Liljenström 1991). With this model we study to what extent its complex dynamics can facilitate an efficient neural information processing. The aim was to capture the essential features of the real system responsible for its dynamic and "cognitive" properties. Since the notion was that the network structure, rather than details of the network units (neurons) were important, a network architecture was used analogous to the three-layered structure of the real cortex, and simple network units corresponding to populations of neurons. The model includes, implicitly or explicitly, adaptation at all relevant time scales, as discussed above.

The network units are modeled with a continuous input-output function, corresponding to the average firing frequency of neural populations. Three different types of network units (cell populations) are organized in three layers. The top layer consists of inhibitory "feedforward interneurons", which receive inputs from an external source (the "olfactory bulb") and from the excitatory "pyramidal cells" in the middle layer. The bottom layer consists of inhibitory "feedback interneurons", receiving inputs only from the excitatory units and projecting back to those. The two sets of inhibitory units are characterized by their different time constants. In addition to the feedback from inhibitory units, the excitatory units receive extensive inputs from each other and from the "olfactory bulb". All connections are modeled with time delays for signal propagation, corresponding to the geometry and fiber characteristics of the real cortex. Figure 8.3a shows the general connectivity in the network.

The time evolution for a network of N neural units is given by a set of coupled non-linear first order differential delay equations for all the N internal states, u. With external input, $I(t)$, characteristic time constant, τ_i, and connection weight w_{ij} between units i and j, separated with a time delay δ_{ij}, we have for each unit activity, u_i,

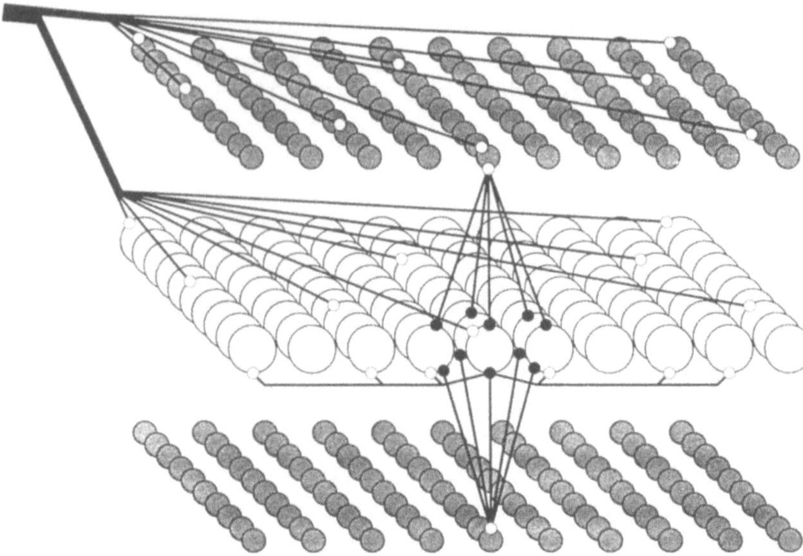

Fig. 8.3a. A schematic figure of the model neural network, whose basic structure is mimicking that of the olfactory cortex. There is one layer of excitatory units, corresponding to populations of pyramidal cells (open circles), in between two layers of inhibitory units, corresponding to two different types of interneurons (light and dark shaded circles, respectively). External input (from "the olfactory bulb") projects onto the two top layers in a fan like fashion.

$$\frac{\mathrm{d}u_i}{\mathrm{d}t} = -\frac{u_i}{\tau_i} + \sum_{j=1}^{N} \eta(t) w_{ij} g_j[u_j(t - \delta_{ij})] + I_i(t) + \xi(t)$$

The input-output function, $g_i(u_i)$, is a continuous sigmoid function, experimentally determined by Freeman (1979):

$$g_i = C \cdot Q_i \left\{ 1 - \exp\left[-\frac{\exp(u_i) - 1}{Q_i} \right] \right\}$$

The gain parameter Q_i determines the slope, threshold and amplitude of the curve for unit i. This gain parameter is associated with the level of arousal, or alternatively, the level of any particular neuromodulator. C is a normalization constant. Neuronal adaptation is implemented as an exponential decay of the output, proportional to the time average of previous output (Liljenström & Hasselmo 1995). With C not a constant, but denoting this adaptation function the input-output relation becomes,

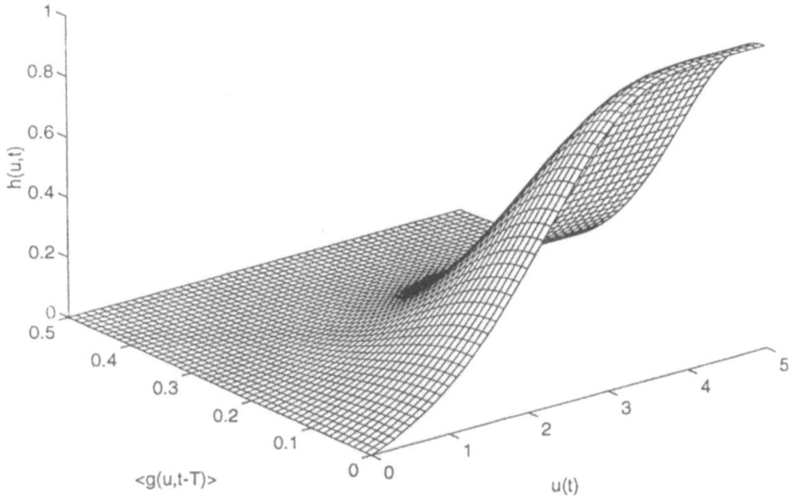

Fig. 8.3b. Unit output dependence on internal activity u, and on previous output history, $\langle g_i(u, t-T)\rangle$, corresponding to neuronal adaptation. The gain parameter is here $Q = 10$, and the adaptation parameter is $a = 5$. (Adopted from Liljenström & Wu 1995)

$$g_{i,\alpha}(u_i) = g_i(u_i)\exp\{-[\alpha\langle g_i(u)\rangle_T]^2\},$$

where $\langle\,\rangle_T$ denotes the time average over the last T ms, and α is an adaptation parameter that is under neuromodulatory (cholinergic) control. The output dependence on the current internal state and on previous activity is shown in Fig. 8.3b.

The connection weights w_{ij} are initially set and constrained by the general connectivity principles that have evolved for the olfactory cortex. To allow for learning, i. e., adaptation, at an intermediate time scale, the weights are incrementally changed according to a learning rule of Hebbian type:

$$\Delta w_{ij} = \nu g_i[u_i(t)] g_j[u_j(t-\delta_{ij})](w_{\max} - w_{ij})$$

The learning rate is denoted by ν, and $w_{\max}$ is the maximum strength of an intrinsic synaptic connection. Spontaneous neuronal activity is added to the system as a Gaussian noise function, $\xi(t)$. In some cases we introduce temporal synaptic noise in the connections, $\eta(t) = \eta_0 + \xi(t)$, which enters as a multiplicative factor in the summation of all synaptic currents. Noise effects on learning and associative memory tasks are studied by continuously changing the level of the additive or multiplicative noise.

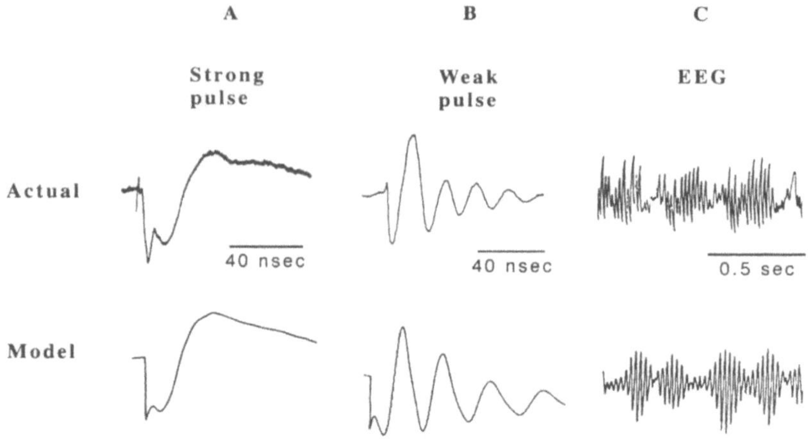

Fig. 8.4. Comparing model results with experimental data from Freeman's lab. The response of a neuron and (excitatory) network unit to A) a strong shock pulse and B) a weak shock pulse. Real and simulated EEG traces are shown under C. (Adopted from Liljenström, 1995)

With computer simulations of this model of the olfactory cortex, we are able to reproduce much of the rich dynamics of the real cortex (Liljenström 1991; Liljenström & Hasselmo 1995). In Fig. 8.4 computer simulations are compared with experimental data from rat olfactory cortex. In particular, input patterns corresponding to different odors result in distributed patterns of primarily oscillatory activity in the network. Such a response is presumed to be related to perception of an odor object. Short input pulses can give rise to waves of activity moving across the simulated cortical surface, similar to what is seen in experimental studies (see Fig. 8.5). The complex dynamics of this model is primarily a result of time constants and network connectivity: feedforward and feedback inhibitory loops, as well as short and long range excitatory-excitatory connections, all modeled with distance-dependent delays. Details at the neuronal level (such as action potentials or cellular compartments) do not seem necessary for the dynamic phenomena studied. However, theta rhythm oscillations can also be produced from neuronal adaptation, which was demonstrated with the inhibitory loops blocked in the simulations (Liljenström & Hasselmo 1995). This is shown in Fig. 8.6.

The model further exhibits associative memory properties similar to those thought to exist in the real cortex, making use of the dynamic be-

Fig. 8.5a. Simulation results showing the dynamics of the three-layered model of the olfactory cortex, following a weak input pulse. Lighter areas correspond to high neuronal activity. (**a**) The activity in the three layers at time $t = 40$ ms

havior previously described (Liljenström 1993, 1995). The oscillating patterns of activity that appear in the network in response to external input patterns are stored as "limit cycle" memory states (see Fig. 8.7), rather than as the "fixed point" memory states typical for most ANN models (e. g. Hopfield 1982). Simulated neuromodulatory effects (of acetylcholine) on the oscillatory behavior and on the memory properties of this system agree well with experimental results and current hypotheses (Liljenström & Hasselmo 1995). In particular, it was shown that the cholinergic suppression of neuronal adaptability and of intrinsic synaptic transmission can facilitate learning by sustaining the oscillatory activity of an input pattern and reducing interference with previously stored patterns (see Fig. 8.8). By changing the gain parameter, corresponding to a change in neuronal excitability, the system can display all types of dynamic behavior: point attractor, limit cycle, or strange attractor behavior (Wu & Liljenström 1994). Simulation results showed that the oscillatory and aperiodic ("transient chaos") dynamic behavior can improve system performance by reducing learning and recall (convergence) time in associative memory tasks.

In addition to the neuromodulatory effects previously described, neuronal and synaptic noise adds to the complex behavior of the system and

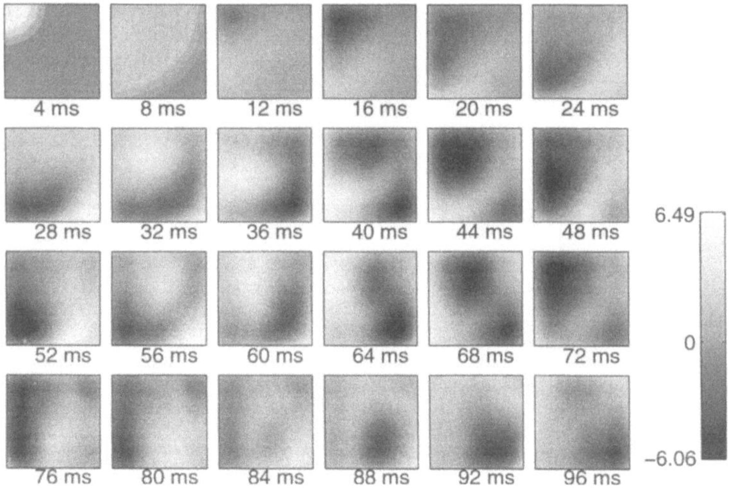

Fig. 8.5b. Each graph displays the activity of the excitatory (middle) layer at time intervals of 4 ms. The short pulse, corresponding to an electric shock pulse given to the lateral olfactory tract (LOT), originates in the upper left corner of the simulated cortical area, and initiates waves of activity that move across the network surface. This simulation which was run on a CM-200 used 64 × 64 network units in each of the three layers, corresponding to an area of 10 × 10 mm of the rat olfactory cortex. (Adopted from Wilhelmsson & Liljenström 1995)

may increase the rate of information processing further. In fact, optimal noise levels can be found for which the recall time reaches a minimum, see Fig. 8.9. Noise can also induce transitions from one (attractor) state to another (Liljenström & Wu 1995).

Most of the simulations were run on regular workstations using one hundred network units in each one of the three layers, i. e., totally three hundred network units. However, with an implementation of our model on a parallel Connection Machine, we can conveniently run much larger simulations. For example, using 3 × 96 × 96 network units with up to 70 million connections in our simulations of the olfactory cortex, we can get more realistic behavior, with a higher spatial resolution and avoiding unwanted boundary effects. We can also easily follow the dynamics of this system for several simulated seconds (Fig. 8.5).

With the type of computer models we have discussed here, attractor neural networks with biologically inspired connectivity, it is possible to capture features of real neural systems and get simulation results that in many respects closely resemble experimental results. In addition to

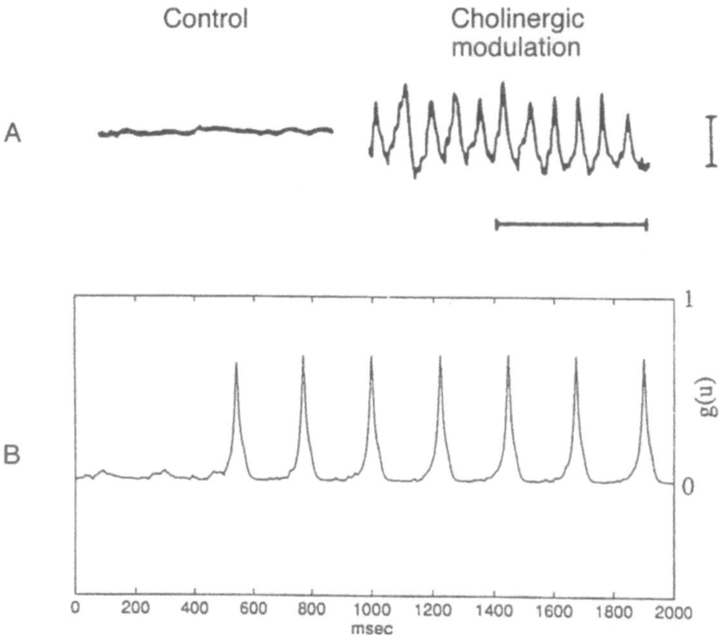

Fig. 8.6. Cholinergic modulation of spontaneous activity in hippocampal neurons. Simulation results with inhibitory synaptic transmission completely suppressed. Increasing the cholinergic input at time $t = 500$ ms induces theta frequency oscillations. (Adopted from Liljenström & Hasselmo, 1995)

the type of simulations presented here, these models can also be used for prediction of temporal sequences, which is currently under study. The models can be useful for finding out which neural properties and network architecture are responsible for the characteristic dynamics of the real system. By exploring the parameter space for, e. g., connection strengths, delays, gains, etc., we can point at plausible solutions to neurobiological problems, and put constraints on the possibilities of the natural system. Although we cannot make any conclusive statements based on the models alone, theories for certain cognitive functions, such as perception, learning, and memory can be tested, and critical experiments can be suggested. Here, we have pointed at the relationship between the network structure and the complex dynamics it displays, using a neural network model of the olfactory cortex. Furthermore, we have linked the dynamics of the system to its function, and shown that

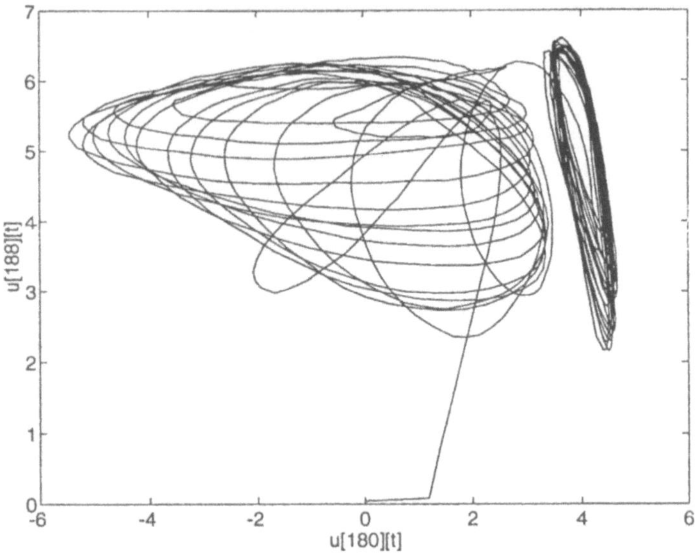

Fig. 8.7. Phase plane plot for one simulated second with recall and learning. The two separate attractors correspond to two separate memory states. A degraded version of a previously stored pattern is given as the input to the network. During the first 500 ms the network tries to recall (converge to) the stored memory state, represented by the large central attractor. When a convergence at the required accuracy fails, the system starts to learn the current input pattern as a novel attractor memory state. (From Liljenström, 1995)

the complex behavior can enhance the performance, by increasing fidelity and rate of information processing.

8.6 Concluding Remarks

In this chapter, we have looked upon mind as a continuous, global process, or set of processes in the brain. In particular, we have regarded certain cognitive functions, such as perception, learning, and memory, as being dependent upon the neurodynamics of large cortical areas. These phenomena are associated with patterns of neural activity, presumably corresponding to internal representations of the environment and of the self. The spatial scale supposed to be relevant for these cognitive functions is in the order of millimeters or larger, and the corresponding temporal scale on the order of one to a hundred milliseconds. All this

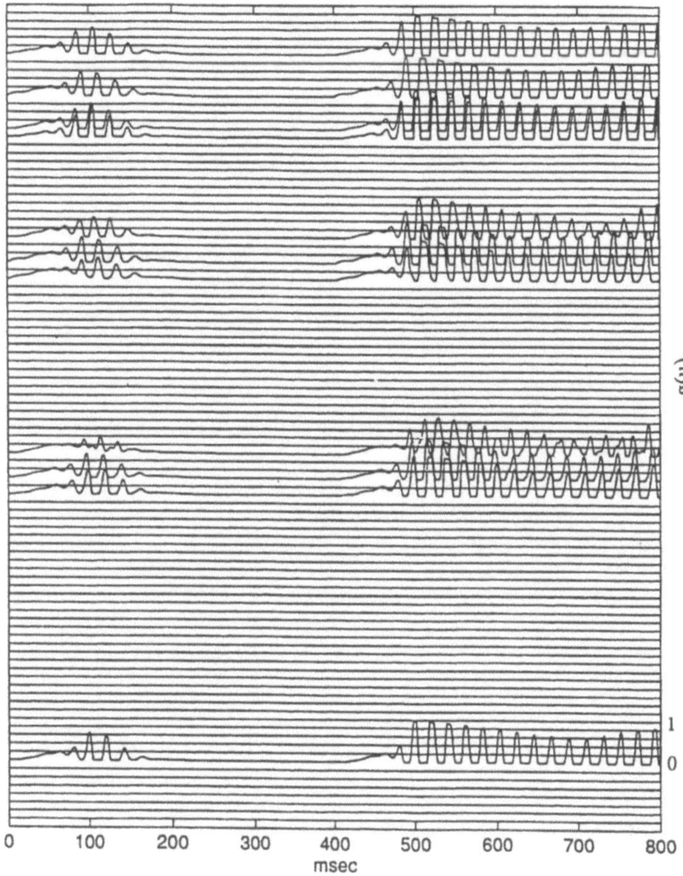

Fig. 8.8. Cholinergic suppression of neuronal adaptation and synaptic transmission set appropriate dynamics for learning. (Adopted from Liljenström & Hasselmo, 1995)

seems to be a result of evolutionary pressure on the nervous system to efficiently deal with and adapt to rapid environmental changes and events.

An interesting aspect of biological systems in general and of neural systems in particular, is their dependence on noise and errors. The "errors", or inaccuracy, in the neural information processing could in fact be the key to human intelligence. This "fuzziness" in our neural processes, which results in some kind of "indeterminism", or at least unpredictability in our actions, does not need to depend on quantum processes (Erdi 1993). It also does not need to involve chaotic processes, in a true mathematical

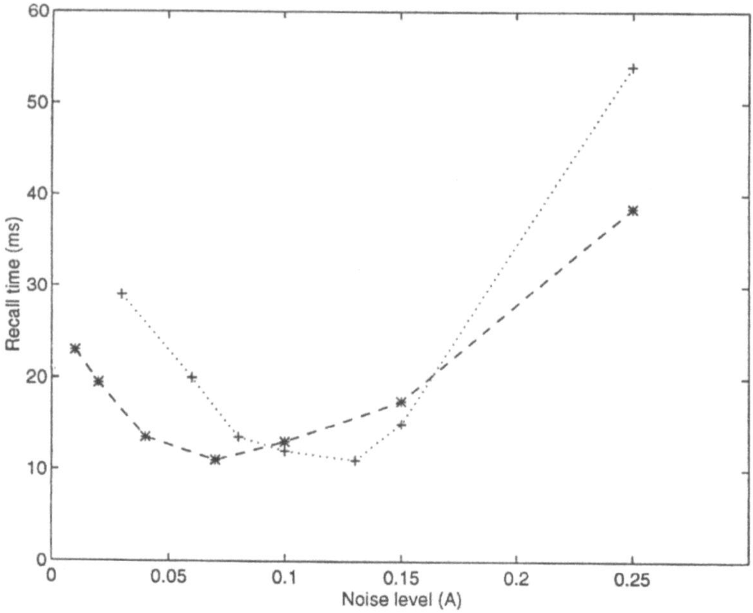

Fig. 8.9. Recall time reaches a minimum for an optimal noise amplitude during learning and recall. For additive noise (o) the optimal noise level is higher than for multiplicative noise (*). The control parameter $Q = 10$ in both cases. Two patterns were stored and a third, degraded pattern was used for recall. (Adopted from Liljenström & Wu 1995)

sense. Chaos, as it is characterized by mathematical means, requires "infinite" time for its development, whereas the neural processes are very unstable and shift from one state to another within a few hundred milliseconds or less. Yet, there is a dynamical region between order and pure randomness that involves a high degree of complexity and which seems characteristic for neural processes. The complex dynamics of the brain can be regulated, e. g., by neuromodulators and perhaps also by noise. By this flexible control the neural system could be put in an appropriate state for the right response-action dependent on the environmental demand.

We have assumed that those cognitive functions that depend on patterns of neural activity can also, at least in principle, be simulated with artificial neural network models. The "connectionist" approach with biological constraints that was described show that computer models can indeed capture some general principles of neural systems, suggesting possible relations between structure and function etc. Although a rather specific part of the brain was chosen as our model system, I expect that many of the model and simulation results can be generalized also to other systems. The olfactory cortex is a phylogenetically old and primitive structure, and many of its properties may have been carried over to other parts of the brain that evolved later, e. g., neocortical structures such as the visual cortex. Even if the structures of various parts of the brain are different, the dynamics may be similar and many of the computational properties could be shared. Hence, it is a hope that simulations like these can shed light on such problems as perception and associative memory, and in their extension, will help us to get an idea of how the brain functions and how it can be related to mental processes in general.

If we accept that mind is a process (or set of processes), mind can never be captured at any one moment. If we could take the brain of a person at any point in his life and measure the level of activity of each neuron and the strength of each synapse, we would not be able to say anything of this person's mind, or even of a single thought. If the *processes* relevant for the cognitive functions of the brain can be realized in any other kind of system, such as an artificial system made of silicon, the same functions will probably appear. The question is whether it is possible to capture and copy all the relevant processes. There may be properties at the cellular or sub-cellular level that are necessary for mental processes and that cannot be incorporated in an artificial system. Yet, it seems as if it is primarily the organization and complexity of the matter that is important, how the neurons are connected and how they interact with each other.

We may in the future be able to describe what neural processes are involved in perception, thinking, reasoning, etc., and crudely simulate these phenomena on a computer, but that does not mean that we have a complete understanding of all the steps and relationships leading from the microscopic level of neurons to the level of conscious experience. With Barlow (1987) I believe that "there would probably be something mysterious about the subjective sensation [of redness] that would remain after one had a complete account of the underlying mechanisms. This 'something' seems likely to lie beyond the range of current scientific approaches."

If one would expect a computer (model) to, in some sense, behave like an animal brain, one would need to incorporate the experience of millions of years of interaction with a changing environment. It has been argued (Skarda & Freeman 1987; Churchland & Sejnowski 1988; Haken

1991) that even if the modeling efforts in the spirit of computational neuroscience will produce extremely complex models that duplicate the performance of the human brain to a significant degree, the models themselves will be hard, or even impossible to understand (in terms of rules or principles that relate the different parts of the system to each other and to other phenomena). This view is expressed by Churchland and Sejnowski (1988):

> Even if we could simulate, synapse for synapse, our entire nervous system, that accomplishment, by itself, would not be the same as understanding how it works. The simulation might be just as much of a mystery as the function of the brain currently is, for it may reveal nothing about the network and systems properties that hold the key to cognitive effects.

MacKay (1980) makes a point in that we cannot understand the human brain (or mind) at any one single level:

> If we do not choose the right logical level at which to give our description, important points may be totally missed. Understanding, as distinct from mere cataloguing, requires the choice of an appropriate level of description for the aspect that we want to understand ... Understanding the human brain can never mean building a completely detailed picture. It has to be a far more modest enterprise, carried out piecemeal by reference to a relatively minute number of sample areas of the whole nervous system ...It is important not to confuse the idea of understanding in this limited sense with the sort of understanding that could lead to complete prediction and manipulation.

In effect, this is the same as to say that we cannot reduce these aspects of (human) consciousness to electrochemical processes or computational algorithms. That would have no meaning to us, even if we claimed we had succeeded. Concepts at one level are in many cases not transferable in a meaningful way to another level. New qualities and properties emerge at each new level in the hierarchical organization of matter, qualities which are irrelevant at lower levels. Such a "holistic" view was also given by Delbrück, who compared mind with quantum reality:

> The mind is not a part of the man-machine but an aspect of its entirety extending through space and time, just as, from the point of view of quantum mechanics, the motion of the electron is an aspect of its entirety that cannot be unambiguously dissected into the complementary properties of position and momentum.

Penrose (1989) seems to bring this analogy further, believing that mind indeed may need some quantum mechanical description. He argues that there are some aspects of mind, or mental phenomena, like understanding and insight, that are non-computable in nature, and thus can never be simulated on a computer. Such phenomena may even require a new physics, new laws and principles that are not mechanistically derivable from lower levels. With this perspective, consciousness seems to be a principle that fundamentally transcends physics, chemistry, and any mechanistic principle of biology.

Regardless of the nature of the mind-brain interaction, our brains' ability to organize the world in an orderly way, relating objects and phenomena to each other, putting singular events and sensory perceptions into a context, brings meaning to the universe. The more we can connect and relate in the world we perceive, the more we realize that the world is one and that there is really no separation between the observer and the observed, between inner and outer world. It is a dynamic whole, in which the parts are mutually dependent and interacting. This brings forth a kind of ultimate meaning, a meaning which fills a person with awe of the immense complexity and beauty of Nature.

References

Ambros-Ingerson, J., Granger, R. & Lynch, G. (1990). "Simulation of paleocortex performs hierarchical clustering", *Science* 247: 1344–1348

Amit, D. J. and Tsodyks, M. V. (1991). "Quantitative study of attractor neural networks retrieving at low spike rates: II. Low-rate retrieval in symmetric networks", *Network* 2: 275–294

Århem, P. & Lindahl, B. I. B. (1997). "On consciousness and spontaneous brain acticity", in this volume

Barlow, H. (1987). "The biological role of consciousness" in *Mindwaves* ed. by C. Blakemore & S. Greenfield, Blackwell, New York, pp. 361–374

Blomberg, C. & Liljenström, H. (1988). "Efficiency in bisynthesis–cellular synergetics" in *Proc. Symp. on Synergetics, Madrid Oct. 12–17, 1987* ed. by M. G. Velarde, World Scientific, Singapore

Bower, J. M. & Koch, C. (1992). "Experimentalists and modelers: can we all just get along?", *Trends in Neurosciences* 15: 458–461

Churchland, P. S. & Sejnowski, T. J. (1988). "Perspectives on cognitive neuroscience", *Science* 242: 741–745

Churchland, P. S. & Sejnowski, T. J. (1992). *The Computational Brain: Models and Methods on the Frontiers of Computational Neuroscience*, MIT Press, Bradford Books

Crick, F. (1984). "Function of the thalamic reticular complex: the searchlight hypothesis", *Proc. Natl. Acad. Sci.* 81: 4586–4590

Crick, F. (1994). *The Astonishing Hypothesis – the Scientific Search for the Soul*, Simon & Schuster, London

Crick, F. & Koch, C. (1990). "Towards a neurobiological theory of consciousness", *Seminars Neurosci.* 2: 263–275

Delbrück, M. (1986). *Mind from Matter?* Blackwell Scientific, Palo Alto, CA

Dennett, D. C. (1991). *Consciousness Explained*, Penguin Books, New York

Downer, J. (1988). *Supersense – Perception in the Animal World*, BBC Books

Eckhorn, R. Bauer, R., Jordan, W., Brosch, M., Kruse, W., Munk, M. & Reitboeck, H. J. (1988). "Coherent oscillations: A mechanism of feature linking in the visual cortex", *Biol. Cybern.* 60: 121–130

Edelmann, G. (1991). *The Remembered Present – A Biological Theory of Conciousness*, Basic Books, New York

Erdi, P., Gröbler, T., Barna, G. & Kaski, K. (1993). "Dynamics of the olfactory bulb: bifurcations, learning, and memory", *Biol. Cybern.* 69: 57–66

Freeman, W. J. (1975). *Mass Action in the Nervous System*, Academic Press, New York

Freeman, W. J. (1979). "Nonlinear gain mediating cortical stimulus-response relations", *Biol. Cybern.* 33: 237–247

Freeman, W. J. (1991). "The physiology of perception", *Sci. Am.* 264: 78–85

Gray, C. M., Konig, P., Engel, A. K. & Singer, W. (1989). "Oscillatory responses in cat visual cortex exhibit inter-columnar synchronization which reflects global stimulus properties", *Nature* 338: 334–337

Haberly, L. B. & Bower, J. M. (1989). "Olfactory cortex: model circuit for study of associative memory", *Trends in Neurosci.* 12: 258–264

Haken, H. (1983). *Synergetics: An Introduction*, Springer-Verlag, New York

Haken, H. (1991). *Synergetic Computers and Cognition*, Springer-Verlag, Berlin

Hebb, D. O. (1949). *The Organization of Behavior*, John Wiley & Son, New York

Heiligenberg, W. (1991). "The neural basis of behavior: a neuroethological view", *Ann. Rev. Neurosci.* 14: 247–267

Hopfield, J. (1982). "Neural networks and physical systems with emergent collective computational abilities". *Proc. Natl. Acad. Sci. USA* 79: 2554–2558

Hopfield, J. (1984). "Neurons with graded response hace collective computational properties like those of two-state neurons". *Proc. Natl. Acad. Sci. USA* 81: 3088–3092

Hopfield, J. J. (1994) "Physics, computation and why biology looks so different", *J. Theor. Biol.* 171: 53–60

Ingvar, D. H. (1985). "'Memory of the future': an essay on the temporal organization of conscious awareness", *Human Neurobiology* 4: 127–136

Lansner, A. & Liljenström, H. (1994). "Computer Models of the Brain – How far can they take us?", *J. Theor. Biol.* 171: 61–73

Levy, W. B. & Baxter, R. A. (1995). "Energy efficient neural codes", *Neural Computation* (in press)

Li, Z. & Hopfield, J. J. (1989). "Modeling the olfactory bulb and its neural oscillatory processings", *Biol. Cybern.* 61: 379–392
Libet, B. (1985). "Unconscious cerebral initiative and the role of conscious will in voluntary action", *Behav. Brain Sc.* 8: 529–566
Liljenström, H. (1987). *On the Efficiency of a Biological System*, Doctoral thesis, The Royal Institute of Technology, Stockholm
Liljenström, H. (1991). "Modeling the dynamics of olfactory cortex using simplified network units and realistic architecture", *Int. J. Neural Syst.* 2: 1–15
Liljenström, H. (1994). "Oscillations and associative memory: brain and model" in *Brain and Mind, Biol. Transactions* 43: 109–123, ed. by R. Cotterill, Danish Academy of Sciences
Liljenström, H. & Hasselmo, M. (1995). "Cholinergic modulation of cortical oscillatory dynamics", *J. Neurophysiol.* 74: 288–297
Liljenström, H. & Wu. X. (1995). "Noise-enhanced performance in a cortical associative memory model", *Intl. J. Neural Systems* 6: 19–29
MacKay, D. (1980). *Brains, Machines & Persons*, W. Collins Sons & Co., London
Malenka, R. C. & Nicoll, R. A. (1993). "NMDA-receptor-dependent synaptic plasticity: multiple forms and mechanisms", *Trends Neurosci.* 16: 521–527
Maturana, H. R. & Varela, F. J. (1992). *The Tree of Knowledge*, Shambala, Boston
Palm, G. (1990). "Cell assemblies as a guideline for brain research", *Concepts in Neuroscience* 1(1): 133–147
Penrose, R. (1989). *The Emperor's New Mind*, Oxford University Press, Oxford
Roland, P. (1993). *Brain Activation*, John Wiley and Sons, New York
Schrödinger, E. (1944). *What is Life?* Cambridge University Press, Cambridge
Skarda C. & Freeman (1987). "How brains make chaos in order to make sense of the world", *Brain and Behavioral Sciences* 10: 161–195
Singer, W. (1990). "Search for coherence: A basic principle of cortical self-organization", *Concepts in Neuroscience* 1: 1–26
Sperry, R. W. (1977). "Forebrain commissurotomy and conscious awareness", *Journal of Medicine and Philosophy* 2: 101–126
Sperry, R. W. (1983). *Science and Moral Priority*, Columbia University Press, New York
Stryer, L. (1981). *Biochemistry*, Freeman & Co., San Francisco
Tsuda (1991). "Chaotic itinerancy as a dynamical basis of hermeneutics in brain and mind," *World Futures* 32: 167–184
van Essen, D. (1985). "Functional organization of primate visual cortex" in *Cerebral Cortex*, vol. 3, Visual Cortex ed. by A. Peters & E. G. Jones, Plenum Press, New York
Wilson, M. A. & Bower, J. M. (1989) "The simulations of large-scale neural networks" in *Methods of Neuronal Modeling: From Synapses to Networks*, ed. by C. Koch & I. Segev, MIT Press, Cambridge
Wu, X. & Liljenström, H. (1994). "Regulating the nonlinear dynamics of olfactory cortex", *Network: Computation in Neural Systems* 5: 47–60

Part IV

The Subjective Level

In earlier blocks of this book we have illuminated the multilevel character of the mind-matter problem by stepping from microscopic levels, as discussed by Hameroff and Conrad & Conrad, to systems and network levels, as discussed by Haken and Liljenström. Here, a reverse path will be followed. The point of departure will be a discussion of the high-level phenomena of human consciousness, including self-consciousness and subjectivity. But the end point will be a discussion on the role of microscopic events for global, macroscopic activity. Thus, we are back at the starting point of our discourse and the block can be seen as closing the book in a natural way.

Although the point of departure for this block is somewhat abstract, it should be noted that both the contributions of Stoerig and of Århem & Lindahl are heavily based on a biological/evolutionary perspective. Mind and consciousness are seen as biological phenomena, brain phenomena or neuronal phenomena, and are approached in biological/evolutionary terms. But it should also be noted that even within this confined subject area, the general theme of a multilevel approach is evident; issues at a macroscopic, brain level being discussed by Stoerig and issues at a cellular, neuronal level by Århem & Lindahl.

The biological approach implies a certain "anti-functionalism". In contrast to the view inherent in some modern forms of functionalism (for an overview, see Churchland, 1988), that specific algorithms are sufficient to describe mental events including subjectivity, it is here argued that matter really matters. Stoerig treats it in the defense of naturalism, i. e., that only living beings possess mentality and consciousness. Århem & Lindahl treat it in the defense of interactionism, i. e., that mental processes in some respect are autonomous and interact with brain processes.

As mentioned, the point of departure for this block is human consciousness. There are many reasons to believe that human self-consciousness is a unique phenomenon. A comparative analysis of biological communication systems provides strong arguments for such a view. But there are reasons to believe that the specific perception of a human self is layered, comprising both more primitive robust layers as well as higher, more complex layers. It is mainly these higher layers that have been studied by traditional psychologists. This many-leveled structure is not always recognized. However, it may be important to make this distinction. It seems today more fruitful, both neurophysiologically and clinically, to study the perception of a primitive self than to study the immensely more complex "Freudian self".

What could be said about non-human forms of self-consciousness? Do such forms exist? Traditionally, self-consciousness is seen as something evolving from the basic capacities of consciousness and awareness. Stoerig argues for a more unorthodox view; self-consciousness and consciousness in general evolve in parallel. But self-consciousness and consciousness are preceded by the self/non-self distinction. The self/non-self distinction is a prerequisite for consciousness, consciousness involves the self as its node.

How is consciousness, including the many layers of self-consciousness, associated with neuronal processes? A strategy to study this complex problem, at this – according to Francis Crick – "pre-Copernican" stage, is to focus on the more primitive forms of consciousness, i. e., perception and awareness. This is the strategy followed by Stoerig in her contribution. More specifically, she concentrates on visual awareness. Her point of departure is the recently recognized phenomenon of blindsight, a phenomenon comprising sensory processing without simultaneous awareness.

This baffling phenomenon offers unique possibilities to analyze which parts and processes are associated with visual awareness. The analysis indicates that out of the many visual areas in the cortex the primary projection area seems to play a fundamental role. A damaged primary area may cause loss of visual awareness. And yet, visual processing may exist. A patient with a damaged primary area may be blind but nevertheless able to follow a moving point of light on a screen with his/her fingers and even discriminate between colors. The view that the primary area plays the essential role for visual awareness is not uncontroversial. It contradicts the more traditional view, recently revived by Crick and Koch (1995), that the visual areas higher up in the processing hierarchy play the main role for visual awareness. Evidently, we do not have the answer to this question yet, but the answer seems not unattainable.

What neuronal processes are associated with visual awareness? This question is, not unexpectedly, even more controversial than the question of the critical structure discussed above. But there are concrete, testable suggestions today. It has recently been proposed that synchronous oscillatory activity in groups of neurons may be essential for awareness (Singer 1994; Crick & Koch 1990). The question of what cellular activity is associated with consciousness is closely related to the much debated question of neural coding, i. e., how neural information is encoded. Is it encoded as the mean frequency of the nerve impulses (rate code) or is it coded in more detailed temporal patterns (temporal code)? Or is neural information encoded in completely other forms, as Hameroff and Conrad & Conrad seem to propose in the first block of this book? And which neurons are performing the critical activity? Are there specific neurons associated with awareness? Even here, there are now concrete suggestions. Koch and Crick (1994) propose that they consist of certain pyramidal cells in the lower layers of cortex. For the first time in the history of science, "realistic" suggestions are today put forward about issues traditionally considered as principally outside science. This is the background against which the discussion of Århem & Lindahl on the role of cellular microevents for brain macroactivity should be seen.

The flexible and creative aspects of thinking make spontaneous brain activity a natural ingredient in the discussion of the consciousness problem. It is clear that spontaneous activity is essential for normal brain function. As Szentagothai (1993) recently pointed out, the static Cartesian reflex paradigm has to give way to more dynamic views on brain activity, including concepts such as spontaneous activity and self-organization. For this reason, the mechanisms of spontaneous activity seem essential for the mind-brain debate.

Århem & Lindahl approach this problem by taking up some experimental findings suggesting that single molecular events may trigger neuronal impulses. This makes the question about the stochastic nature of these single molecular events important. Can they be described within the framework of deterministic, but perhaps chaotic models (in the mathematical sense), or do we need indeterministic Markov (or other) models? The latter case raises the possibility, already touched upon by Penrose, Hameroff, and Conrad & Conrad, that quantum mechanical processes may be essential for brain function. The philosophical consequences of the answer to these questions may be far-reaching, touching on such questions as free will and rationality. However, whatever the outcome of such an analysis will be, the classical philosophical mind-brain problem remains unanswered. Does further detailed knowledge about the molecular cellular mechanisms associated with consciousness fully explain consciousness? Are certain neuronal processes perfectly parallel with mental processes, or are they, in some sense, separate and interactive? This question seems to require an analysis at another level.

However, the discussions in the present book seem to suggest that the mind-brain debate could be improved by replacing the traditional controversy between materialism and immaterialism with a more concrete and fertile discussion on the computability of neural and mental processes. It is probably not too bold a conjecture to assume that the somewhat limited aim of explaining visual and other sensory awareness is not infinitely far away. But reaching such a goal only marks the beginning of attempts to understand the higher-level problems of rational decisions, the "self" or free will, i. e., the classical domains of what we see as expressions of human consciousness.

References

Churchland, P. M. (1988). *Matter and Consciousness: A Contemporary Introduction to the Philosophy of Mind*, 2nd ed., MIT Press, Cambridge, MA

Crick, F. & Koch, C. (1990). "Towards a neurobiological theory of consciousness", *Seminars in the Neurosciences*, Vol. 2, pp. 263–275

Crick, F. & Koch, C. (1995). "Are we aware of neural activity in primary visual cortex?", *Nature* 375: 121–123.

Koch, C. & Crick, F. (1994). "Some further ideas regarding the neuronal basis of awareness" in *Large Scale Neuronal Theories of the Brain* ed. by C. Koch & J. L. Davis, MIT Press, Cambridge, MA, pp. 93–110

Singer, W. (1994). "Putative functions of temporal correlations in neurocortical processing" in *Large Scale neuronal Theories of the Brain* ed. by C. Koch & J. L. Davis, MIT Press, Cambridge, MA, pp. 201–238

Szentagothai, J. (1993). "Self-organization: the basic principle of neural functions", *Theor. Med.* 14: 101–116

9. Consciousness and the Matter of Perception

Petra Stoerig

9.1 Functionalism or Matter Does not Matter

"The real question is not whether machines think, but whether men do" (Skinner 1969, p. 288).

When I read the title of the symposium, *Matter Matters*, I thought, ah, an anti-functionalist workshop. I interpreted it like this because the thesis that matter matters is contrary to the modern functionalist one, which says that it does not matter what kind of matter is the actual basis realizing the set of operations, and that any system doing so will by definition have the functional properties we do, which include mentality (Fodor 1975; Dennett 1991). I have a few problems with this approach.

The term functionalism as it is mainly used now, for instance by Fodor and Dennett, differs from the much older use in which the mind was conceived of as a function of the brain, as rotation is a function of wheel (Uttal 1978). The older use was closer to materialist theories that treated mind as a product of the brain (Priestley 1777), sometimes in disrespectful analogy to urine being a product of the kidneys (Vogt 1855). Whereas these older concepts were monistic in a strict sense, the newer versions with their impact on the independence of the algorithm or set of operations from the kind of matter used for its realization – a living being, a computer, a set of Coca Cola cans – introduce a peculiar dualism between the algorithm and matter.

What would be a convincing way to determine whether an artifact that uses a set of algorithms possesses conscious mentality? The behavior that it displays is the first means that comes to mind. But is it sufficient? Does conscious mentality not involve something in addition to overt behavior, namely this famous inner aspect that allows one to pore over one's past and future failures? Even subtle overtly observable behavioral nuances that the artifact's makers could introduce – for instance for it to refrain from responding under certain conditions, to remain immobile as if in contemplation of a piece of art, or to act as if it found something funny and someone especially likable, following him or her around – would not prove that it does experience what we associate with these behaviors – voluntary restraint in a social context, aesthetic contemplation, humor, attraction[1].

The problem is that we do not have a convincing test. But, just assuming that we had an artifact whose behavior was indistinguishable from that of a conscious living being, what would we then have to conclude? That it has mentality and consciousness which account for the behavior? That it has mentality and consciousness but that this does not causally influence the behavior? That it does not have mentality, and therefore that mentality is not necessary to display purposeful behavior? And consequently, that mentality and consciousness do not play a causal role in the behavior of living beings? Or what? The questions about mind, matter, and consciousness would still be open to deliberation.

Furthermore, people and other animals (at least chimpanzees) would have a strong tendency to ascribe mentality to something that behaves as if it possessed it. Children ascribe intentionality to self-propelled objects (Premack 1993), an adaptation, one may assume, to our not having encountered anything other than living beings that behaves purposefully in the course of our evolution. After all, self-propelled artifacts are a rather recent innovation not known to be fabricated by non-human animals, and the few early examples fashioned by man in earlier centuries caused great wonder, probably without tuning our cognitive equipment to distinctions between artifactual purposefulness (caused by the outside will of the artificer) and self-motivated intention. Therefore, behavior alone, important though it is, does not suffice to assess conscious mentality.

9.2 Naturalism, Life, and Self

In view of the problems I have with the functionalist approach, I restrict my interest in the questions of mentality and consciousness to their presence and implementation in living beings, defending the naturalist thesis that living beings possess mentality and consciousness. My reason is the intrinsic link I see between life, mentality, and consciousness. This link follows from two major characteristics of living beings. The first is the drive for preservation and replication/procreation. Even very simple organisms try to protect themselves from noxious stimuli in their environment, and try to get as much of a good thing as they can. As soon as you have life, no matter how that came into this world, you also have needs, which you have to satisfy in order to stay alive. To protect yourself, to survive, after all the first function of reason according to Whitehead (1929), you have to be able to distinguish between self and non-self, between what belongs to your organism and what does not, between yourself and your environment, between inside (me) and outside (not-me). Several people have pointed this out, among them Edelman (1989) and Dennett (1991) who wrote: "As soon as something gets into the business of self-preservation, boundaries become important, for if you are setting out to preserve yourself, you don't want to squander effort trying to preserve the whole world: you draw the line. You become, in a word, *selfish*" (p. 174).

An organism does not need to be very complex to show these two characteristics; even a paramecium shows the appropriate behavior. What about plants? Flanagan, in his book *Consciousness Reconsidered* (1992), argues "If one is a mere photosynthesizer and ground-nutrient absorber, the self/non-self distinction is not one that the system needs to be sensitive to" (p. 50). Flanagan seems to suggest that plants, or more generally, non-moving beings, do not need this distinction, and Humphrey (1992) also stresses the importance of the ability to move, the learning of what you can and what you cannot move voluntarily, for this distinction. I agree that at least under normal circumstances the ability to move means that you are able to make this distinction. However, I should argue that an inability to move does not preclude the distinction between self and non-self. One perceives parts of one's body as being part of oneself even when they cannot be moved, provided one can feel them. As long as one has sentience, and distinguishes between a pin being stuck into one as opposed to a pin being stuck into one's bed, one's neighbor, some non-self, one can develop and entertain the distinction between self and non-self. I assume that without sentience this distinction is impossible, and I am not aware of good grounds to deny either to plants[2] and other immobile organisms per se. However, I am not aware of good grounds to attribute selfishness to artifacts.

I suggest that the self/non-self distinction is a prerequisite for consciousness; that consciousness – however much complexity may be required for it to appear – involves the self as its node. The self is primary biologically as well as epistemologically, because every sentient organism represents its reality – however coarsely – in relation to it-self, and stores information in relation to it-self[3]. This does not mean that self-consciousness precedes consciousness. Instead, it implies that conscious representations of an organism's environment and those of itself develop hand in hand, because the distinction requires a representation of both. This suggestion is thus in contrast to the common assumption that consciousness eventually culminates in self-consciousness which is often reserved for the human primate.

If the self is the node or premise of consciousness, does every organism capable of the self/non-self distinction possess consciousness? It is possible, but it is equally possible that the organism needs to be able to represent events and conditions inside and outside of itself in a manner that allows information to be processed or "reflected" between input and output.

9.3 Input, Output, and In Between

Behavior is by itself insufficient to assess the presence or absence of conscious mentality. One telling example is a condition called *coma vigilante* in which behavior is rendered impossible by a small but critical lesion in the pontine area of the brain. Patients who recover from this state can give lucid accounts of their experiences during the coma (see Gillett 1988). Despite its shortcomings as an indicator of consciousness, behavior is nevertheless methodologically necessary for its assessment, and is shown in response to internal or external stimulation. In the simplest case, the reflex, the input and the output are immediately linked[4]. This may be the only form of behavior of which a very simple organism with no or a very low level of functional differentiation is capable. Serving the organism's basic needs, reflexes still persist in higher organisms, and in some cases are mediated by just one central neuron linking the receptor and the effector level in monosynaptic reflex arcs. The input, whether it is a hot stove touched by a child's hand, a swill of water aimed at an *Aplysia*'s gill, or light falling onto a cat's pupil, is translated into the appropriate output pattern – withdrawal and constriction, respectively – without intervening contemplation. Reflexes may be elicited when the organism is asleep. Conscious volition is not involved[5].

At the other end of behavioral complexity are optional actions and reactions where the organism may respond in a number of different ways to a situation. Whether an organism capable of such behavior acknowledges the presence of a fellow depends not only on its ability to sense (see, smell, hear, feel, echolocate, or whatever) its presence, but also on the state in which it is (active or lazy), and on previous experiences with this or other conspecifics. A dog may wag his tail when he meets one fellow dog, put the same piece of equipment firmly between his hindlegs when he meets another, and do neither when he relaxes in the sun and a third one comes by. The translation between input and output thus is subject to a variety of influences, as the inner state (activation level), previous experiences (beware of Dobermans), and the external context (he's on a leash) are taken into account. At this stage, behavior alone is insufficient to describe the organism's activity; contemplating a mathematical problem, conceptualizing philosophical ideas, day-dreaming, scheming, and playing with possible outcomes of various courses of action need not appear different from napping in the sun. These behaviors do involve conscious mentality (although the mind may unconsciously work on problems and possible solutions during sleep once the problems have been sufficiently internalized), and require elaborate intermediary nets.

Intermediate nets are also needed to mediate the vast number of behaviors that range between these two poles of behavioral complexity. They include instinctive behaviors like the egg rolling of a goose, or the aggressive attacks of a stickleback, which are subject to deception from mischievously curious experimenters who substitute the competitor with a chimera, the egg with a beer bottle. They also include fixed-action patterns and perceptual-motor skills such as walking, swimming, feeding, and spinning a spider's web. They are considerably more complex behaviorally than reflexes, and therefore require much more elaborate neuronal mediation between receptor and effector patterns to account for the behaviors' well-adapted complexity. Nevertheless, it is by no means certain that these programs *need* to involve conscious experiences. The automatism which characterizes certain instinctive behaviors has often been cited to argue against conscious deliberation, and the fact that one can walk quite skillfully whilst asleep shows that consciousness need not be involved in the execution (although possibly in the acquisition) of fixed-action patterns[6].

The intermediary net required for any complex behavior has undergone continuous elaboration in mammalian and avian evolution. Man is estimated to have 10^{11} or 10^{12} neurons, of which less than 1000 million are receptor and motor neurons, leaving more than 99 000 000 000 neurons for the intermediary net. Somewhere in its phylogenetic development, a stage is reached where conscious phenomena – which then co-exist more or less peacefully with the already established reflexes, fixed-action programs, and instinctive behaviors – first appear.

9.4 Specialization Within the Intermediary Net

In man, the organism whose mentality we are most familiar with, the general functional specialization within the CNS is studied with methods which include the correlation of behavior and the registration of neuronal events, for instance the recording of potentials evoked by certain tasks, or by imaging the brain during various types of overt behavior, like speech, reading, counting, or moving a limb, or during covert behaviors such as calculating silently, or imagining characters from a TV household series. Information about the representation of function is also gathered in the damaged brain, in that cognitive and behavioral deficits are assessed and correlated with the brain damage, its nature, location, and extent, often in combination with functional neuroanatomy.

These methods have taught us that the frontal lobes contain the motor cortex, the speech production area, aspects of social and planning behavior; that the parietal lobes contain the somatosensory cortex; that the temporal lobes contain auditory cortex as well as language areas, and that the occipital lobe (plus portions of parietal and temporal lobes) contain the visual cortex. What we do not know, but may be able to learn through more refined experiments and better technologies, are the properties that neurons or neuronal networks must possess to produce conscious phenomenal perceptions, or conscious thoughts and decisions.

Neurons are distinguished on the basis of many criteria:

- Their place in the brain: whether they are cortical or subcortical; to which functional system they belong; if they're cortical, whether they're in primary sensory cortex, secondary sensory cortex, multisensory cortex, motor cortex, or whether they form parts of cortex that serve neither sensory nor motor function such as the prefrontal lobes; their place in the cortical layers, of which there are six in neocortex.
- Their connectivity: what number and kinds of inputs they receive and where they project, whether the input is pre- or post-synaptic; whether it comes from cortical or subcortical neurons.

- Their discharge properties – sluggish or bursting, transient or sustained.
- Their morphology: in the cortex, the main distinction between neurons is whether they are pyramidal or non-pyramidal, the non-pyramidal cells being subdivided into a number of morphologically distinct classes (stellar, basket). Morphological distinctions already play a role at the sensory organ; in the retina, morphological cell classes differ also in their functional properties and in their projection areas.
- The neurotransmitters they use, whether these are excitatory or inhibitory.

To address at least a fraction of these possibilities (always keeping in mind the to me unlikely possibility that all neurons within the intermediary net are equipotential in their ability to realize conscious phenomena and experiences), I shall use the visual system as an example. The deficits which are caused through lesions at its various stages may shed some light on the locus and connections needed to subserve conscious phenomenal vision.

9.5 Blindness, Blindsight, and Phenomenal Vision

Any sense depends on its sensory organs. If you lose both eyes, the functional consequence is that you are blind. Blindness means that you cannot process any visual information, that you cannot respond to it, and in cases where you have never had functional eyes because you were born blind, it supposedly entails that you don't have visual imagery or visual dreams either. Blindness is the consequence of never having had functional eyes, or of losing the eyes, or the retinae, or the optic nerve, although the ability to have seen before may make a difference with respect to visual dreams and imagery, because a previously developed ability to excite the visual cortex from an extra-retinal source may remain, at least for some time[7].

The optic nerves that transmit the visual information from the eye into the brain have about 10 target areas. In functional terms this means that if you suffer a lesion that spares the fibers traveling to some of these extrageniculate destinations, you may retain certain reflexive visual responses, i. e., your pupils may still dilate or constrict depending on the amount of light falling onto your eyes, or you may be able to make reflexive eye movements. These reflexive responses may not be entirely normal because generally several parts of the system participate in their mediation (even the pupil response seems to have a cortical contribution), and because the various parts of the system are intimately connected in a way such that destruction of one nucleus can inhibit function in another even though the other does still get its retinal input (Sprague 1966; Wallace et al. 1989). Nevertheless, the rule of thumb is that with lesions at this stage you retain reflexive function in response to visual information, but no voluntary control, and you do not see in any experiential fashion.

Once the lesion is at the next stage, destroying the optic radiations and/or the primary visual cortex, homonymous visual field defects result. Even from absolute field defects, a large variety of responses to visual information can be elicited. In addition to pupil reflexes, skin conductance changes, and avoidance responses to visual stimuli, they include saccadic and manual localization, target and motion detection as well as orientation and wavelength discrimination (see Weiskrantz 1986, 1991; Cowey & Stoerig 1991; Stoerig & Cowey 1993a, for recent reviews). These responses can be voluntarily initiated by the patients, and are statistically significant, sometimes even 100% correct. Nevertheless the patients generally claim they are only guessing to please the experimenter. At best, they say they feel something happening, which occurs when large or vigorously moving stimuli are used.

This of course is the phenomenon of blindsight (Weiskrantz et al. 1974) which has recently found its way with increasing frequency into philosophical discussions (e. g., Dennett 1991; Hyman 1991; Flanagan 1992; Nelkin 1993; Stoerig and Brandt 1993). It is characterized by statistically significant processing of all kinds of visual information in the absence of a phenomenal visual experience on the part of the patient. This lack includes an inability to visually imagine scenes or objects (Farah et al. 1992[8]). As testable cases with bilateral cortical blindness are rare, it is unknown whether blindsight patients have visual dreams, which would be important for the claim that you eventually lose phenomenal visual perception of any kind with such a lesion.

Whereas destruction of the entire belt of extrastriate visual cortical areas also causes blindness (Bodis-Wollner et al. 1977), damage confined to parts of extrastriate cortex produces deficits such as cortical color blindness, stereo blindness, and motion blindness, depending which of the specialized cortical areas are destroyed by the lesion. Contrary to a V1 lesion where all visual information has lost the capacity to produce a conscious phenomenal perception, these extrastriate lesions prevent only a certain type of visual information from being consciously represented. As in blindsight, however, this type of information can be used implicitly by the patients, for instance to infer the position of cues and borders (Heywood et al. 1991). These patients have phenomenal visual perceptions, which lack a visual modality.

Once the lesion is past the striate cortex, and does not destroy the entire circumstriate cortex, the patients have phenomenal conscious vision. Although visual modalities may be lost due to localized extrastriate cortical damage, although the phenomenal percept may be "stripped of its meaning" (Lissauer 1890), the patient unable to recognize a face, a set of keys, a house as his own, it is always a conscious visual perception that remains – even if it is quite useless in severe agnosias (see Farah 1990, for a recent review). Striate cortex therefore appears to be the critical stage for phenomenal visual perception.

9.6 The Matter of Perception

What is special about the primary visual cortex? One possibility is that its position in the sensory processing hierarchy, with its providing the main input to the higher visual cortical areas, is responsible for its particular importance. This possibility is very much in line with the currently predominant thinking about hierarchical processing. The extrageniculo-striate cortical inputs which permit visual responsivity in at least some of the extrastriate cortical areas after inactivation of striate cortex (see Bullier et al. 1993, for a recent review) would then be insufficient to cause phenomenal vision[9]. As destruction of the optic radiation and destruction of the striate cortex produce similar losses of phenomenal vision, the geniculo-striate input to the higher areas is indeed necessary. Is it sufficient?

It is also possible that the importance of the primary visual cortex is caused not only by its distributing retino-geniculate information into a hierarchical cortical processing system, but also by its being the recipient of the information processed by the higher areas and sent in the opposite direction to the primary one. The function of the massive back-projections within the cortical network, still poorly understood, has recently been addressed in the somatosensory system. Cauller and Kulics (1991) trained monkeys to perform a touch discrimination task, in which two stimuli were applied to the monkey's hand, and he had to decide which of the two was the stronger. The experimenters then made this discrimination more and more difficult until a point was reached where the monkey's decisions were quite independent of the actual difference between the stimuli. At this stage, they recorded the evoked potentials from the monkey's cortex, and found one potential, called N1, that allowed them to predict the monkey's response – remember that their knowing which one was the stronger stimulus didn't do. This potential is attributed to the projections from higher somatosensory cortical areas back to the primary cortex, and disappears under anaesthesia and during dreamless sleep (Arezzo et al. 1981; Cauller & Kulics 1988).

It follows that the importance of the primary cortex for conscious sensation may not be solely due to its distributing incoming information to the higher areas, but in addition to its receiving the processed output from the higher areas. If this were the correct explanation, a preferably transient inactivation of the projections from the higher cortical areas to the primary one should affect phenomenal sensation. A comparison between the effect of this inactivation and one that is restricted to the outgoing, feed-forward connections could address this issue[10].

The evidence from the somatosensory system might well hold for the visual system as well. Phenomena characterized by a loss of phenomenal conscious experience in the presence of established information processing have been described for sensory modalities other than vision; deaf hearing (Michel 1990) and unfeeling touch (Paillard et al. 1983) demonstrate that different sensory systems realize similar organizational principles.

My line of argument implies that the neuronal basis of phenomenal consciousness may eventually be scientifically defined. With an interdisciplinary approach that encompasses neuropsychology, -physiology, -anatomy, -pathology, -psychiatry, with the study of natural and experimental deficits and disturbances of these and other aspects of consciousness, we may come to a reasonably good understanding of the processes realizing conscious experience. We may need techniques that we do not yet possess, but I do not see why we should not go on trying to distinguish the in-fact and the in-principle-consciously-represented processes from those neuronal events that are not consciously accessible. After all, the brain, in addition to realizing conscious representations, partakes in the regulation of breathing, mediates hormone production, and processes incoming information from the senses and outgoing patterns of motor activity. These latter functions affect mentality, change the mood, the emotions, the plans, and eventually even the personality of the organism possessing the brain – but then so does a rock falling onto someone's foot, a meal being eaten, or a pill suppressing the menstrual cycle. None of these are mental. The mental part is in between input and output, depending on both, and it interacts with the brain's other functions. Where we can assess it, it requires an intermediary net with properties we may eventually be able to understand[11].

Notes

[1] See Searle (1992) on the relationship of mind and behavior.
[2] See Turlings et al. (1993) for a recent example of self-protective behavior in corn seedlings.
[3] Gennaro (1992) argues that episodic memory needs to entail self-consciousness.
[4] I do not wish to imply that the reflex is the paradigmatic neuronal function; I regard it as the simplest form of behavior, which is consistent with results showing that neuronal "activity starts in the very early stages of development well before receptors and afferent input become functional" (Szentágothai 1993, p. 101).
[5] If one is conscious anyway, certain reflexive responses (the knee jerk reflex, but not the pupil reflex) get consciously represented after having been initiated or executed. Like other autonomous responses, reflexes can – with persistent training and often with external feed-back – be brought under voluntary control. This was described by Thomas Mann whose protagonist Felix Krull succeeds to control his pupil reflexes: "Anfangs gerieten unter den inneren Anstrengungen, die mir den Schweiß austrieben und mich die Farbe wechseln ließen, meine Pupillen nur in ein unregelmäßiges Flackern; später aber hatte ich es tatsächlich in meiner Gewalt, sie sich zu winzigen Pünktchen verengern oder zu großen, schwarz spiegelnden Kreisen sich ausdehnen zu lassen, und die Genugtuung, die dieser Erfolg mir gewährte, war fast schreckhafter Art und von einem Schauer vor den Geheimnissen der menschlichen Natur begleitet" (1960, p. 274).
[6] Both consciousness and unconsciousness cover a continuum. One can be more or less conscious, depending on the state of alertness, and more or less unconscious, depending on the depth of sleep or coma. During dreams one can experience vivid and extraordinary adventures and emotions. These one experiences as happening to/being watched by oneself (no matter what form that self may take during the dream), showing that the self is present in that state. In addition, the dreamt experiences are both phenomenal and qualitative, and often if not always (contrary to common appearances) meaningful and intentional. Dreamt experiences are not usually introspectable during the dream, although it happens that one is aware of the fact that one is only dreaming, but it is possible to recall dreams in the awake state, and to then consciously contemplate them. Whether phenomenality, intentionality,

and introspectability need to be present for an experience to be classified as conscious is only one of the open questions. (See Nelkin 1993, for a discussion of these three components of consciousness.)

[7] Note that I talk about absolute, not legal blindness. This is important, and likely to be the reason for the discrepancies in the published literature. Legally blind people may have a diffuse light and even form perception which entails functional connections between the eyes and the visual cortex, and probably explains reports of visual imagery in the congenitally blind, such as Burkhard (1965). (See also Zimler & Keenan 1983; Hollins 1985.) It may also explain the differences observed after surgical interventions such as removing the cause of blindness in operations of congenital cataract (Latta 1904) or in corneal graft implantation (Gregory & Wallace 1963).

[8] Apart from the study of Farah et al. (1992) who report a clear decrease in the horizontal extent of imaged objects after unilateral occipital lobectomy, there is very little known about visual imagery in cortical blindness. Ideally, one would study a patient who has lost his or her striate cortex only, without additional damage to extrastriate cortical areas, and who is known to have had visual imagery before the lesion. This is necessary because people appear to differ substantially in their ability to "see" imagined scenes. Electrophysiological (Farah et al. 1988) and blood flow (Goldenberg et al. 1987) studies implicate the striate cortex in visual imagery. As the images have to be evoked from memory, not from the retina, disturbances due to lesions in structures that are not strictly visual may also disturb imagery (see Guariglia et al. 1993 for a recent report of unilateral neglect for visual imagery attributed to damage in the right frontal lobe).

[9] The visual hallucinations in the hemianopic field which a fraction of patients perceive following damage to the occipital lobe (e. g., Kölmel 1985) would then be explained by (possibly excitotoxic) processes simulating striate cortical input. Their disappearance within a matter of days, and very occasional reappearance in periods of extreme stress, is consistent with denervation hypersensitivity.

[10] To test these suggestions in a monkey, one needs to know whether he/she suffers the same loss of phenomenal conscious perception the patients describe following striate cortical lesions. To this end, one needs means to disentangle the monkey's phenomenal perceptions from his/her behavioral responses indicating the residual visual functions which can be demonstrated after striate cortical removal in monkeys (e. g., Klüver 1941; Humphrey 1974). Alan Cowey and I are presently trying to develop an approach that may at least take us a step along the way (Stoerig & Cowey 1993b).

[11] What constitutes understanding is a big question. A definition of the neuronal processes mediating conscious experiences does not necessarily yield an understanding (in the strong sense) of how a subjective conscious experience is realized by a neuronal process with this set of properties. But then I for one am equally unable to predict that water appears liquid and transparent from knowing that its composition is H_2O. The quality of water can only be empirically correlated with its molecular structure, and without anyone perceiving the quality, it is unpredictable. In sci-

ence, we explain the quality which is the first given by quantitative-structural means. Qualities seem quite impervious to current scientific explanations in that they cannot be constructed from the quantitative description alone. In this sense, it is on a non-scientific or strong understanding of understanding that McGinn (1991) argues against the possibility of a neuroscientific solution to "the problem of consciousness".

Acknowledgement. My thanks go to Alan Cowey, Norton Nelkin, and Shimon Ullman, who contributed valuable information, comments, and critique. The work was supported by the Deutsche Forschungsgemeinschaft (Sto 206-4/1).

References

Arezzo, J. C., Vaughn, H. C. & Legatt, A. D. (1981). "Topography and intracranial sources of somatosensory evoked potentials in the monkey: II. Cortical components. Electroencephalography", *Clinical Neurophysiology* 28: 1–18

Bodis Wollner, I., Atkin A., Raab, E. & Wolkstein, M. (1977). "Visual association cortex and vision in man: pattern-evoked occipital potentials in a blind boy", *Science* 198: 629–631

Bullier, J., Girard, P. & Salin, P.-A. (1993). "The role of area 17 in the transfer of information to extrastriate visual cortex" in *Primary Visual Cortex in Primates* ed. by A.Peters & K.S.Rockland, Cerebral Cortex 10, Plenum Press, New York, pp. 301–330

Burkhard, U. (1965). "Von der Farbvorstellung blindgeborener Menschen". Schweizerisches Jahrbuch *"Die Ernte"*, pp. 40–46

Cauller, L. J. & Kulics, A. T. (1988). "A comparison of awake and sleeping cortical states by analysis of the somatosensory-evoked respose of postcentral area 1 in rhesus monkey", *Experimental Brain Research* 72: 584–592

Cauller, L. J. & Kulics, A. T. (1991). "The neural basis of the behaviorally relevant N1 component of the somatosensory-evoked potential in awake monkeys: Evidence that backward cortical projections signal conscious touch sensation", *Experimental Brain Research* 84: 607–619

Cowey, A. & Stoerig, P. (1991). "The neurobiology of blindsight", *Trends in Neuroscience* 14: 140–145

Dennett, D. (1991). *Consciousness Explained*, Little Brown, New York

Edelman, G. M. (1989). *The remembered present: A biological theory of consciousness*, Basic Books, New York

Farah, M. J., Peronnet, F., Gonon, M. A. & Giard, M. H. (1988). "Electrophysiological evidence for a shared representational medium for visual images and percepts", *Journal of Experimental Psychology: General* 117: 248–257

Farah M. J. (1990) *Visual Agnosia. Disorders of Object Recognition and What They Tell Us about Normal Vision*, MIT Press, Cambridge, MA

Farah, M., Soso, M. J. & Dasheiff, R. M. (1992). "The visual angle of the mind's eye before and after unilateral occipital lobectomy", *Journal of Experimental Psychology: Human Perception and Performance* 18: 241–246

Flanagan, O. J. (1992). *Consciousness Reconsidered*, MIT Press, Cambridge, MA

Fodor, J. (1975). *The Language of Thought*, Crowell, New York

Gennaro, R. J. (1992). "Consciousness, self-consciousness and episodic memory", *Philosophical Psychology* 5: 333–347

Gillett, G. R. (1988). "Consciousness and brain function", *Philosophical Psychology* 1: 327–341

Goldenberg, G., Podreka, I., Steiner, M. & Willmes, K. (1987). "Patterns of regional cerebral blood flow related to memorizing of high and low imagery words: An emission computer tomography study", *Neuropsychologia* 25: 473–486

Gregory, R. L. & Wallace, J. G. (1968). *Recovery from Early Blindness. A Case Study*, Cambridge.

Guariglia, C., Padovani, A., Pantano, P. & Pizzamiglio, L. (1993). "Unilateral neglect restricted to visual imagery", *Nature* 364: 235–237

Heywood, C. A., Cowey, A. & Newcombe, F. (1991). "Chromatic discrimination in a cortically colour blind observer", *European Journal of Neuroscience* 3: 802–812

Hollins, M. (1985). "Styles of mental imagery in blind adults", *Neuropsychologia* 23: 561–566

Humphrey, N. K. (1974). "Vision in a monkey without striate cortex: a case study", *Perception* 3: 241–255

Humphrey, N. (1992). *A Natural History of the Mind*, Simon & Schuster, New York.

Hyman, J. (1991). "Visual experience and blindsight" in *Investigating Psychology: Sciences of the Mind after Wittgenstein* ed. by J. Hyman, Routledge, London, pp. 166–200

Klüver, H. (1941). "Visual functions after removal of the occipital lobes", *Journal of Psychology* 1: 23–45

Kölmel, H. W. (1985). "Complex visual hallucinations in the hemianopic field", *Journal of Neurology, Neurosurgery, and Psychiatry* 48: 29–38

Latta, R. (1904). "Notes on a case of successful operation for congenital cataract in an adult", *British Journal of Psychology* 1: 135–150

Lissauer, H. (1890). "Ein Fall von Seelenblindheit nebst einem Beitrage zur Theorie derselben", *Archiv für Psychiatrie und Nervenkrankheiten* 21: 222–270

Mann, Th. (1960). "Die Bekenntnisse des Hochstaplers Felix Krull", *Gesammelte Werke in 13 Bänden, vol. VII*, S. Fischer, Frankfurt

McGinn, C. (1991). *The Problem of Consciousness*, Blackwell, Oxford

Michel, F. (1990). *Hemi-anacusia is usually unknown to the patient*, Poster presented at the Russell Trust/Wellcome Trust Symposium, St. Andrews, Sept. 1990

Nelkin, N. (1993). "What is consciousness?", *Philosophy of Science* 60: 419–434

Paillard, J., Michel, F. & Stelmach, G. (1983). "Localization without content: a tactile analogue of 'blindsight'", *Archives of Neurology* 40: 548–551

Premack, D. (1993). "Prolegomenon to evolution of cognition" in *Exploring Brain Functions. Models in Neuroscience* ed. by T. A. Poggio & D. A. Glaser, John Wiley & Sons, Chichester, New York, pp. 269–290

Priestley, J. (1976). *Disquisitions Relating to Matter and Spirit*, Garland (1st ed. 1777) London, New York

Searle, J. R. (1992). *The Rediscovery of the Mind*, MIT Press, Cambridge, MA

Skinner, B. F. (1969). *Contingencies of Reinforcement*, Appleton-Century-Crofts, New York

Sprague, J. M. (1966). "Interaction of cortex and superior colliculus in mediation of visually guided behavior in the cat", *Science* 153: 1544–1547

Stoerig, P. & Brandt, S. (1993). "The visual system and levels of perception: properties of neuromental organization", *Theoretical Medicine* 14: 117–135

Stoerig, P. & Cowey, A. (1993a). "Blindsight and perceptual consciousness" in *Functional Anatomy of the Human Visual Cortex* ed. by B. Gulyás, D. Ottoson & P. E. Roland, Pergamon Press, Oxford, pp. 181–193

Stoerig, P. & Cowey, A. (1993b). "Perception and phenomenal consciousness", *Abstracts 25th Annual Meeting of the European Brain and Behaviour Society*

Szentágothai, J. (1993). "Self-organization: the basic principle of neural functions", *Theoretical Medicine* 14: 101–111

Turlings, T. C. J., McCall, P. J., Alborn, H. T. & Tumlinson, J. H. (1993). "An elicitor in caterpillar oral secretions that induces corn seedlings to emit chemical signals attractive to parasitic wasps", *Journal of Chemical Ecology* 19: 411–425

Uttal, W. R. (1978). *The Psychology of Mind*, Erlbaum, Hillsdale, NJ

Vogt, C. (18553) *Köhlerglaube und Wissenschaft. Eine Streitschrift gegen Hofrat Ludwig Wagner*, Gießen

Wallace, S. F., Rosenquist, A. C. & Sprague, J. M. (1989). "Recovery from cortical blindness mediated by destruction of nontectotectal fibers in the commissure of the superior colliculus in the cat", *Journal of Comparative Neurology* 284: 429–450

Weiskrantz, L., Warrington, E. K., Sanders, M. D. & Marshall, J. (1974). "Visual capacity in the hemianopic field following a restricted occipital ablation", *Brain* 97: 709–728

Weiskrantz, L. (1986). *Blindsight. A Case Study and Implications*, Oxford University Press, Oxford

Weiskrantz, L. (1990). "Outlooks for blindsight. Explicit methodologies for implicit processes", *Proceedings of the Royal Society*, London, B 239: 247–278

Whitehead, A. N. (1929). *The Function of Reason*, Princeton University Press, Princeton, NJ

Zimler, J. & Keenan, J. M. (1983). "Imagery in the congenitally blind: How visual are visual images?", *Journal of Experimental Psychology: Learning, Memory, and Cognition* 9: 269–282

10. On Consciousness and Spontaneous Brain Activity

Peter Århem and B.I.B. Lindahl

10.1 Introduction

In a biological perspective the central mind-body problem is how and why the brain creates consciousness. Although the solution to this problem may be extremely complex or even unattainable, the problem itself is quite comprehensible. It can be subdivided into a number of questions: Why have brains with consciousness survived and evolved? Which brain processes are crucial for consciousness? Which parts of the brain are associated with consciousness? And which processes and parts of the brain are crucial for the specific human consciousness?

The problem of consciousness is indeed not only biological, but clearly transdisciplinary (see Århem & Lindahl 1993a,b). However, without an evolutionary and a neurophysiological approach to the problem, there is little hope for progress (Crick & Koch 1990; Edelman 1992; Blomberg et al. 1994; see also Århem 1987). On a long view it is not very fruitful to explain away consciousness, as in the instrumentalist tradition of Ryle (1949), as if mental terms do not refer to anything involved in causing behavior (Dennett 1991) or as if they do not refer to anything at all (Churchland 1986) (see Lindahl & Århem 1994). Consciousness is an outstanding feature of the world we live in and has to be explained.

As Crick and Koch points out, the lack of an exact explicit definition of consciousness need not be an obstacle in the analysis.

> Everyone has a rough idea of what is meant by consciousness. We feel that it is better to avoid a precise definition of consciousness because of the dangers of premature definition. Until we understand the problem much better, any attempt at a formal definition is likely to be either misleading or overly restrictive, or both. (Crick & Koch 1990)

In this chapter we will argue, as others have done, that an evolutionary perspective suggests an interactionistic solution to the mind-brain problem (Sect. 10.2). The prevalent reluctance to accept interactionism seems to be due to a combination of a rather dogmatic ontological approach, focusing mainly on the problem of what "stuff" mind may consist of, and a tendency in the scientific community today to adopt an instrumentalistic view of theories, treating scientific theories exclusively as predictive tools and not as propositions about reality (Sect. 10.3). The reluctance to accept an interactionistic theory of the mind-brain relation is reflected in a common tendency not to take consciousness seriously in neuroscience and cognitive studies. However, there are neurophysiological studies partly dealing with the problem of consciousness (Sect. 10.4). In this chapter some experimental findings of relevance in this context will be discussed. The findings concern a requirement that seems fundamental to an interactionistic theory of the mind-brain relation, namely spontaneous brain activity (Sect. 10.5). Finally, a realist explanation of the mind-brain relation, a hypothesis recently proposed by Popper (Popper et al. 1993), will be commented upon (Sect. 10.6). Many of the points argued in the following are extensions of ideas developed in an earlier paper (Lindahl & Århem 1994).

10.2 The Evolution of Consciousness

The importance of a neurophysiological approach does not entail that it is enough to simply state that *mind is the brain*. This is often done, not only in neurophysiological circles (Changeux 1986; Crick 1994), but also in philosophical circles with a more psychological bent (Dennett 1991). Popper puts the problem succinctly:

> There are also those who simply say that the mind is an activity of the brain, and who leave it at that. Not much can be said against this, as far as it goes. But it does not go far enough: the question arises whether the mental activities of the brain are just part of its many physical activities or whether there is an important difference; and if so, what we can say about the difference. (Popper & Eccles 1977, p. 96).

A recurrent answer to this question is that mental activities (the conscious as well as the unconscious) are only parts of the brain's many physical activities; that there is no important difference; and that the mental activities therefore can be fully accounted for in terms of present day physics and chemistry. This is the basic assumption of many of the most influential theories today, the identity theory, functionalism, and eliminative materialism (see, e. g., Feigl 1967; Churchland 1986; Churchland 1988). Many of these theories are virtually variants of *psycho-physical paral-*

lelism. However, one major problem with parallelism, pointed out already by William James (1879), is that it fails to explain how consciousness has evolved in nature. In order to evolve, consciousness must have been naturally selected and this in turn indicates that it may have played a causal role. This argument is discussed in Richards (1987, chapter 9) and Lindahl (1997). (See also Lindahl & Århem 1994.)

The power of James' argument is evident also when applied to current computer theories of mind. One may conceive of cognitive systems with very advanced information processing without assuming a consciousness. At the same time we are acutely aware of our own consciousness, of our subjective experiences. So, if human beings really could do without consciousness, why has evolution bestowed on us, and probably other species, this remarkable trait? The most likely explanation is, of course, that brains with consciousness are better adapted to life on earth than brains without; that consciousness has a survival value. This seems to imply that mental states have a causal influence on brain states and *vice versa* (see Popper 1972; Popper & Eccles 1977). This is the answer of *psycho-physical interactionism* to the mind-brain problem.

Although James' argument does not conclusively refute the idea that consciousness is an epiphenomenon without causal influence on the brain, as proposed by Thomas Huxley (1893), the evolutionary aspect of consciousness is a strong argument against most parallelistic and epiphenomenalistic ideas, and a strong argument in favor of an interactionistic solution of the mind-brain problem (Århem 1987; Lindahl & Århem 1994; Lindahl 1997). The evolutionary argument has been forcefully developed by for instance Popper (Popper & Eccles 1977) and it constitutes the background of the present chapter.

10.3 The Reluctance to Accept Interactionism

Contrary to what is often claimed (see for instance Crick 1994), parallelistic ideas about the mind-body relation have a long history and are by far the most influential today. Why, then, is there such a reluctance to accept interactionistic ideas? As we noted earlier, it seems partly to have to do with a tendency to adopt an outmoded materialistic ontology, and partly with prevalent instrumentalistic views on scientific theories. This will be briefly commented upon below.

The Materialistic Ontology

A number of arguments against interactionism can be found in the literature. Most of them are based on the idea that an immaterial entity (the mind) influencing a material entity (the body) and conversely is not compatible with the notion of causality in natural science. This may be called *the mechanistic argument*. Originally this argument was based on a Cartesian concept of matter, as something extended and impenetrable, and a Cartesian notion of causality, confined to action by contact. Also other, more modern materialistic arguments are to some extent based on the Cartesian notions of matter and causality. This is all the more interesting, considering the fear in philosophical circles of being associated with Cartesian ideas (see Searle 1992). As we have pointed out in an earlier discussion, much of the current debate for and against interactionism restates the well-known Cartesian dilemma: the attempt to unite into one theory a dualistic ontology – the distinction between an immaterial mind and a material body – and a materialistic mechanistic notion of causality (Lindahl & Århem 1994).

With the development of the Newtonian concept of action at a distance, and with the modern concept of force, introduced by Faraday and Maxwell, and partly anticipated by Leibniz, Kant, and Boscovich (see Jammer 1957; Capek 1961; Hesse 1961; Popper 1984), the Cartesian dilemma is no longer relevant (Popper & Eccles 1977; Popper et al. 1993; Lindahl & Århem 1994). Popper bases his interactionistic theory on this development. (This will be further discussed in Sect. 10.6.)

Another common objection against interactionism is *the thermodynamic argument*, that an interaction between something immaterial and something material would violate the laws of thermodynamics. Popper (1984) argues against the idea that the influence of something nonphysical upon something physical would violate the first law of thermodynamics: (a) that the brain is a detector-amplifier providing its own energy (see Eccles 1953), making the law of energy conservation irrelevant for the mind-brain issue, (b) that there may exist mental forms of energy convertible into electrochemical forms, and (c) that there may exist nonenergetic influences (energy-less pilot waves) upon energetic processes (energy-carrying waves). Popper has previously suggested that the first law may be only statistically valid (Popper & Eccles 1977). Popper also argues that the second law of thermodynamics is constantly violated by Brownian motion according to the Einstein-Smoluchowsky interpretation and thus loses its power also in the mind-brain debate (Popper 1984; see also 1976). (For a discussion of the thermodynamic argument, see also Lindahl & Århem 1994, 1996b and Beck 1996.)

The Instrumentalistic View on Theories

There is an interesting parallel between the current discussion about the nature of mind and the controversy in physics at the turn of the century about the nature of unobservable entities, for example mass and force. The similarities between these two debates are striking both in the ways unobservable entities have been understood to exist, or have been denied to exist, and in the way these understandings and denials have been justified. The parallel concerns the problem how theories about unobservable entities should be understood. In both debates, realist as well as instrumentalist interpretations can be discerned. The choice of interpretation decisively affects the direction of future research. A realist interpretation, which understands the assertions about a certain unobservable entity to be true or false, and this regardless of what has been or can be experienced, leaves open the possibility of further investigating the nature of this entity. An instrumentalist interpretation, by contrast, precludes this possibility, by denying that the assertions are anything but more or less useful, or useless, theoretical constructs for making predictions.

The instrumentalistic view on theories can be traced back as far as Cardinal Bellarmino's objection to Galileo's interpretation of the Copernican System, and back to Bishop Berkeley's criticism of Newton's dynamics (Popper 1953b, 1956). In these cases the basis for the instrumentalistic view was religious. Instrumentalism in our times, however, is usually based on a phenomenalistic theory of reference implying that reference can only be made to what has been or can be (at the time in question) intersubjectively experienced. A very influential proponent of this form of instrumentalism, within both physics and philosophy, was Ernst Mach (Blackmore 1972).

Although Mach lost the famous battle against Boltzmann and other realists about the physical existence of atoms – a fact Mach never really conceded (Blackmore 1972) – Mach's phenomenalistic theory of reference has continued to exert great influence on how theories about unobservable entities are viewed in physics. Instrumentalism has become an orthodoxy.

... physics concerns making models, that we make models of the world about us, to help us to relate one type of observation to another. And we either have good models or less good models. And that there is no such thing as a 'real world' in the sense of something which exists 'out there' to which our models are mere approximations. (Davies & Brown 1986, p. 124)

As an example of the negative influence of instrumentalism on physics Popper (1956) mentions Bohr's principle of complementarity. Popper points out that this principle has remained completely sterile within physics: "In twenty-seven years [in 1956] it has produced nothing except some philosophical discussions, and some arguments for the confounding of critics (especially Einstein)". Mach's theory of reference was also a primary source of inspiration for the early logical positivists.

In contrast to instrumentalism, realism claims that assertions about unobservables (like mind) refers to objectively existing entities. Putnam (1979) describes realism in the following way:

The statements of science are in my view either true or false ... and their truth or falsity does not consist in their being highly derived ways of describing regularities in human experience. Reality is not part of the human mind; rather the human mind is a part [...] of reality. (Putnam 1979, p. vii)

Some realist attempts to understand the mind-brain problem will be discussed below.

10.4 Neurophysiological Studies of Consciousness

Instrumentalistic views are evident even in the neurosciences. This is reflected in the tendency to neglect the problem of consciousness, in spite of its key role in the theories of mind (for other examples, see Århem 1996). As Searle points out:

In one way or another, all other mental notions – such as intentionality, subjectivity, mental causation, intelligence, etc. – can only be fully understood as *mental* by way of their relations to consciousness ... (Searle 1992, p. 84)

However, attempts have been made within neurophysiology to understand the problem of consciousness. Most of them aim at correlating consciousness with certain parts of the brain. Of this type are the well-known investigations on functional lateralization, pioneered by Sperry (Sperry et al. 1969). The interpretation of these studies with reference to the problem of consciousness is complex and gives room for many alternatives. Eccles

(1973) and MacKay (1966) suggest that the studies support the view that higher forms of consciousness are associated exclusively with activity in the left hemisphere. Sperry (1977) contends that surgical disconnection of the hemispheres results in two conscious minds or selves within the one cranium. In contrast, Puccetti (1977) argues that studies support the view that higher forms of consciousness are represented in both hemispheres even in intact brains; thus suggesting two separate minds. Similarly, studies of the visual system have given rise to various interpretations. Especially the phenomenon of blindsight has recently been analyzed in this context. Based on such studies, Stoerig and Brandt (1993) call attention to the backprojections to primary areas of visual cortex (mainly area V1) as one possible explanation of conscious perception.

Other neurophysiological studies aim at correlating certain brain activity with consciousness. An example of this kind of studies is the proposal by Crick and Koch (1990) that the semi-synchronous activity of neurons in the 40–70 Hz range, reported in studies of the visual cortex (Gray & Singer 1989), is associated with conscious visual experience. To this second kind belongs also the studies by Libet (1973; Libet et al. 1979) on the relation between neural activity and perception, indicating that subjectively experienced time need not follow the objectively measured time; findings that have been suggested to support an interactionistic view on the mind-brain relation (Popper & Eccles 1977; Penrose 1989; for a criticism of this interpretation, see Churchland 1981; Dennett 1991; Hesslow 1994). Similar interpretations and consequent controversies have followed upon the findings of Libet (1985) on brain activity associated with voluntary movements.

As indicated, these studies have been extensively discussed. Below, attention will be given to another area of neurophysiological studies that seems relevant in the present context, but which has up till now been relatively little explored. It concerns the spontaneous activity of the brain. On both theoretical and experimental grounds, it is today evident that random neuronal activity is a fundamental functional principle in the central nervous system. Theoretically, it is clear that a description of nerve activity based on discrete stochastic ion channel populations, but not a description based on the more commonly used continuous deterministic Hodgkin–Huxley equations, implies spontaneous activity (Strassberg & DeFelice 1993). Experimentally, it has been shown that spontaneous activity is vital for establishing functional nerve connections during ontogeny (see Thompson 1994). As pointed out by Szentágothai (1993) the dominating influence of the Cartesian reflex paradigm for analyzing the nervous system has to give way to a new line of thinking, possibly including concepts such as self-organization and autopoiesis (Maturana & Varela 1980; Érdi 1993). There are reasons to believe that the discussion on spontaneous activity also is central for understanding the mind-brain issue (Lindahl & Århem 1994). In the following some recent findings concerning the mechanisms of such activity will be outlined.

10.5 Brain Mechanisms of Spontaneous Activity

Interactionistic hypotheses seem to require spontaneous brain activity. They are based on the assumption that physical states are open to influences from mental states, not describable in physical terms. That the physical world is indeterministic might not be controversial today, due to the prevalent interpretation of quantum mechanics. However, it is somewhat controversial that the macroactivity of the brain is indeterministic. On the other hand, modern chaos theory has increased interest in indeterministic macroprocesses caused by amplification of microindeterministic processes (see Érdi 1993, and Liljenström, this volume). In fact, Popper has suggested that the well-known *all-or-nothing principle* for the nerve impulse (i. e., that the amplitude of the action potential is independent of stimulus intensity above the threshold value) may refer to such an amplification process:

> The all-or-nothing principle of the firing of nerves may indeed be interpreted as a mechanism which would allow arbitrarily small fluctuations to have macroscopic effects. (Popper & Eccles 1977, p. 541)

In the following this suggestion by Popper will be discussed in light of empirical evidence for a process that amplifies random molecular fluctuations to nerve impulses in the living brain. What mechanisms cause random spontaneous activity in the nervous system? Most discussions of this problem have focused on system properties, in a rather imprecise way. Much less effort has been devoted to considerations of intrinsic neuronal mechanisms, i. e., mechanisms that cause spontaneous activity in a single neuron (see Llinas 1988). This will be done here. While the mechanism of regular, pacemaker activity is well studied and known in some detail (see Jack et al. 1975; Berridge & Rapp 1979), the mechanism of random spontaneous activity is much less studied. Several possibilities are conceivable, however. Before discussing these it is necessary to outline briefly the mechanisms of nerve activity.

The Mechanisms of Nerve Activity

The basic electrical event in the nervous system is the action potential, the time course of the electrical potential difference over the membrane. This forms a wave of electricity that is conducted along the nerve fiber. The underlying mechanism was unveiled by Hodgkin and Huxley in their classic voltage clamp studies on the giant squid axon in the early 1950s (see Hodgkin 1964), and followed up by Frankenhaeuser (Dodge & Frankenhaeuser 1958) for the more complex vertebrate nerve fiber. The cause of the action potential was found to be time and potential dependent Na^+ and K^+ currents through the membrane. During the 1980s, due to the development of the new patch clamp technique (Neher & Sakmann 1976), it became evident that the ion currents passed through water-filled pores in the membrane (see Hille 1992). Further studies revealed that the pores were constituted by specific membrane proteins, ion channels. Several ion channel types have now been classified and cloned with molecular biological methods. The present picture is both diverging and converging. The number of channels described increases continuously – concerning mammalian voltage-gated channels at least six Na channel types and 16 K channel types have today been identified. At the same time a fundamental building plan begins to emerge. This plan depicts the standard channel as a protein (or an aggregate of proteins) of about 280 kilodaltons, consisting of four membrane-spanning domains with a selectivity filter close to the extracellular side and a gating mechanism at the intracellular side (Fig. 10.1).

The gating mechanism, the mechanism of channel opening and closing, is still unknown. The most widely held view is that the process is a time-homogenous Markov process with discrete channel states, meaning that the future of the system exclusively depends on the present state and not on its prehistory (Colquhoun & Hawkes 1995; Horn 1984). Common to almost all descriptions is the assumption that random events form the basis for the activity (for an opposite view see Liebovitch & Toth 1991). Some of these have located the basic randomness at a quantum-mechanical level (Donald 1990).

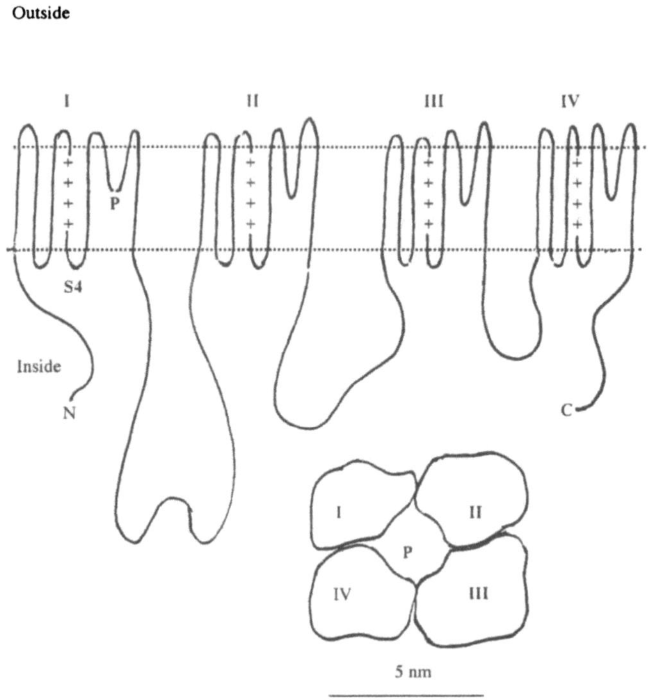

Fig. 10.1. Topology of voltage-gated Na channel. The four clusters of membrane spanning segments are labelled I–IV. The six membrane spanning segments S1–S6 in each domain form a-helices. The charged S4 segment is indicated. The loop forming the walls of the channel pore is labelled P. The packing of the domains is shown schematically below (after an electron micrograph of a K channel, Li et al. 1994)

Explanations of Spontaneous Activity

One tentative explanation of spontaneous brain activity would evidently be to base it directly on random openings or closings of single ion channels. For such a phenomenon to occur, however, certain rather unique properties of the neuron are necessary. Normally thousands of channels are necessary for exciting a neuron at rest. In order to induce a sufficient potential change by a single channel current, either the single current or the membrane resistance must be unusually large (according to Ohm's law). In some secretory cell types (Fenwick et al. 1982) and in olfactory sensory neurons (Lynch & Barry 1989) the membrane resistance seems sufficiently large and single channel induced activity has been recorded. Recently such activity has also been successfully demonstrated in certain brain cells (Johansson & Århem 1994). The background of the study was the following.

Until recently it has not been experimentally possible to record from small neurons (i. e., neurons with a diameter < 10 μm; the diameter for neurons normally used is about 100 μm). However, the development of patch clamp methods (Neher & Sakmann 1976) made this, as well as recordings from single ion channels, feasible. Since small neurons in certain parts of the CNS constitute a considerable fraction of the total number of neurons, and thus, in accordance with an evolutionary perspective, can be expected to play important functional roles, Johansson and Århem set out to clarify the electrophysiological properties of these. The study focused on small neurons from hippocampus, a phylogenetically old part of the cortex, known to be crucial for memory forming mechanisms and other cognitive functions. Somewhat unexpectedly, these cells were found to show at least two interesting features, not earlier described for central neurons. Both features are relevant for understanding information processing in the nervous system. As already mentioned, one was that the action potentials could be induced by single channel events. The other was that the amplitude of the action potential is graded with the stimulus pulse, thus deviating from the generally accepted *all-or-nothing principle* (although the action potentials are still regenerative.)

Figure 10.2 suggests evidence for the conclusion that one form of spontaneous activity depends on openings of single channels. At present it is not known which channel type is responsible for this form of spontaneous activity, but this is not decisive for the present discussion. The study suggests that the neurons studied and the single channel induced activity may be important for the origin of the spontaneous activity in areas of the brain associated with consciousness.

As pointed out in Lindahl & Århem (1994), this suggestion fulfills two fundamental requirements of Popper's earlier mentioned proposal, that the *all-or-nothing principle* allows arbitrarily small fluctuations to have macroscopic effects: the suggestion is based on (i) fluctuations, possibly caused by quantum mechanical events (see the previous section), which are (ii) amplified to macroevents (regenerative action potentials). It is relevant here to note that such macroevents, single action potentials, have been shown to cause conscious experience in humans (Ochoa & Torebjörk 1983).

The detailed mechanism of channel gating is still not known. However, it is reasonable to assume that the crucial trigger structure of the channel is

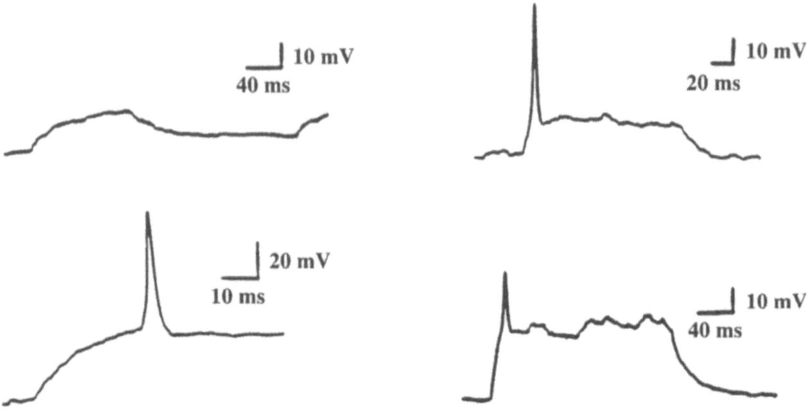

Fig. 10.2. Spontaneous potential activity of cultured hippocampal neurons, measured with patch-clamp technique, whole-cell configuration. One trace (upper left) shows exclusively a plateau potential, the other three include regenerative action potentials. The roughly exponential rise, the plateau phase, and the roughly exponential decay of the plateau potential suggest that it is induced by opening of single ion channels (see Johansson & Århem 1994)

only a fraction of an ion channel-protein. In voltage gated channels, such as Na and K channels, a certain membrane spanning segment consisting of about 20 amino acids and consequently having a weight of about 2×10^{-21} g has been assumed to be critical for the gating. The mass of the structure is important for the discussion about the relevance of quantum-mechanical explanations. In this context 2×10^{-21} g may seem large. However, the triggering event *per se* may depend on a much smaller structure. An interesting proposal has been made by Lee (1992), who suggests that a single electron transfer between two amino acids (a tyrosine and a tryptophan residue), located in the pore wall, may be the critical opening event in voltage gated channels.

This explanation of intrinsic spontaneous activity in a neuron seems more promising than the presently more discussed alternatives, based on the probabilistic nature of synaptic activity. ("Probabilistic" here refers to the fact that a packet of transmitter molecules may be released spontaneously from the presynaptic neuron in a all-or-nothing fashion but also that a nerve impulse may release a packet of transmitter molecules with a probability below 1.0.) Such nerve impulse induced probabilistic release (with probabilities between 0.5 and 0.01) has recently been observed in several synapse types (Jack et al. 1981; Cunnane & Stjärne

1982; Korn & Faber 1987). Eccles (1986) has suggested that a specific microstructure of the synaptic apparatus (the presynaptic grid) is crucial for the probabilistic release of transmitter packets. One weakness with this hypothesis, as explanation of spontaneous activity, compared to the ion channel hypothesis proposed above, is that the critical structure (part of the presynaptic vesicle, i. e., 10^{-18} g) is many orders of magnitude larger than the critical structure for triggering a channel opening (10^{-21} g or much less). Another weakness of this hypothesis, is that it requires joint action of a large number of synapses on the same neuron; random release of a single packet of transmitter molecules is as a rule not sufficient to trigger action potentials in central neurons.

These hypotheses, both the ion-channel hypothesis and that based on the probabilistic synaptic activity, are still only tentative and not experimentally corroborated as explanations of spontaneous brain activity *in vivo*. Other tentative explanations may easily be found. However, the discussed hypotheses are based on experimental findings, and the adoption of an evolutionary perspective forces us to speculate about their functional role.

10.6 The Nature of Mind: Popper's Hypothesis

The existence of spontaneous brain activity *per se*, however, is of course not enough for explaining consciousness. As stated by Popper:

> ... quantum indeterminacy does not seem to help us; for the amplification of something like, say, radioactive disintegration processes would not lead to human action or even animal action, but only to random movements. (Popper 1978)

Some form of selection process seems necessary. Popper suggests a parallel with the selection of mutations in biological evolution (later also done by for instance Calvin 1987, and Edelman 1992):

> Thus, what I am here suggesting is that we might conceive of the openness of World 1 [the world of physical objects] to World 2 [the world of subjective experiences] somewhat on the lines of the impact of selection pressures on mutations. The mutations themselves can be considered as quantum effects; as fluctuations. Such fluctuations may occur, for example, in the brain. In the brain there may at first arise purely probabilistic or chaotic changes, and some of these fluctuations may be purposefully selected in the light of World 3 [the world of theoretical, i. e., logical, conceptual, etc., products of the human mind] in a way similar to that in which natural selection quasi-purposefully selects mutations ... The action of mind on the brain

may consist in allowing certain fluctuations to lead to the firing of neurons while others would merely lead to a slight rise in the temperature of the brain. (Popper & Eccles 1977, pp. 540–541)

Also other theorists discuss the possibility of quantum-mechanical processes being involved. Penrose (1989) suggests that the states of awareness are coherent quantum states of the brain. Beck and Eccles (1992) suggest that the critical structure for these processes are the presynaptic grids in the synapse region of certain cells in the cortex. Donald (1990) suggests the Na channel for this role.

Popper has recently proposed a new hypothesis (Popper et al. 1993) within his interactionistic mind-brain theory (Popper 1953a, 1955, 1972, 1994; Popper & Eccles 1977). This hypothesis is analyzed and discussed in Lindahl and Århem, 1994 (see also Lindahl & Århem 1996a, 1996b; Libet 1996, 1997; and Beck 1996). Popper's point of departure is the similarity between mind and physical forces. He points out that both minds and forces are: (i) located, (ii) unextended, (iii) incorporeal, (iv) capable of acting on bodies, (v) dependent upon body, (vi) capable of being influenced by bodies, (vii) intensities, and (viii) extended through a span of time (Popper et al. 1993, p. 168).

Popper not only calls attention to this analogy but even seems to contend that at least the unconscious mind may be *identical* with an electromagnetic field (Popper et al. 1993; see also Lindahl & Århem 1994).

The mind-force analogy is not new. It goes back to an analysis by Leibniz (see Popper & Eccles 1977). Leibniz identified a part of mind (*conatus*, endeavor, will) with a physical force. Even earlier, Gilbert "had compared the interaction between magnetic force and a lodestone to that between soul and body" (Watkins 1974). What is new with Popper's hypothesis, however, is that it gives clues for experimental tests. By suggesting that electromagnetic fields may play a role in the mind-brain relation, the hypothesis implies that the three-dimensional structure of the brain is relevant to the nature of mind. In this Popper's hypothesis is compatible with the neurophysiological approaches mentioned earlier in this chapter (Crick & Koch 1990; Stoerig & Brandt 1993; Johansson & Århem 1994). However, Popper's hypothesis challenges most of the recent computer theories of mind, such as many of those discussed by Churchland and Sejnowsky (1992) and Dennett (1991); in these and in other functionalistic hypotheses the relevance of the exact three-dimensional structure of the brain in causing consciousness is not considered.

10.7 Conclusion

To sum up, in this chapter we have argued that an evolutionary approach suggests an interactionistic interpretation of the mind-brain relation. The reluctance to accept interactionism was suggested to be due to a combination of an outmoded ontology and an instrumentalistic view on theories in the scientific community. This seems to preclude attempts to construct realistic hypotheses about consciousness that can be tested at least partially.

We have emphasized the importance of a realist view on theories of mind and on the mind-brain problem. We have argued that such a view on the problem of consciousness, based on an evolutionary and neurophysiological approach, is central for the solution of the mind-brain problem. We have stressed the futility of continuing the materialism-immaterialism debate.

Some, albeit fragmentary, experimental attempts to clarify aspects of the problem of consciousness were discussed. Attention was called to Popper's recent and thought-provoking hypothesis about the relation between consciousness and the brain.

Acknowledgement. The experimental parts of this chapter were supported by the Swedish Medical Research Council (Project number 6552). This chapter was also financially supported by the Swedish Council for Research in the Humanities and Social Sciences (F375/94).

References

Århem, P. (1987). "On artificial intelligence and neurophysiology. Two necessary questions" in *Real Brains, Artificial Minds* ed. by J. L. Casti & A. Karlqvist, North-Holland/Elsevier, New York, pp. 181–196

Århem, P. (1996). "Vertical information flow in the brain: On neuronal micro events and consciousness", *BioSystems* 38: 191–198

Århem, P. & Lindahl, B. I. B. (1993a). "Neuroscience and the problem of consciousness: Theoretical and empirical approaches. An introduction", *Theor. Med.* 14: 77–88

Århem, P. & Lindahl, B. I. B., eds. (1993b). "Neuroscience and the problem of consciousness: Theoretical and empirical approaches", *Theor. Med.* 14(2)

Beck, F. (1996). "Mind-brain interaction: Comments on an article by B. I. B. Lindahl & P. Århem", *J. Theor. Biol.* 180: 87–89

Beck, F. & Eccles, J. C. (1992). "Quantum aspects of brain activity and the role of consciousness", *Proc. Natnl. Acad. Sci. USA* 89: 11357–11361

Berridge, M. J. & Rapp, P. E. (1979). "A comparative survey of the function, mechanism and control of cellular oscillators", J. *Expl. Biol.* 81: 217–279

Blackmore, J. T. (1972). *Ernst Mach: His Work, Life, and Influence*, University of California Press, Berkeley, CA

Blomberg, C., Liljenström, H., Lindahl, B. I. B. & Århem, P. eds. (1994). "Mind and matter: Essays from biology, physics and philosophy", *J. Theor. Biol.* 171(1)

Calvin, W. H. (1987). "The brain as a Darwin Machine", *Nature* 330: 33–34

Capek, M. (1961). *The Philosophical Impact of Contemporary Physics*, D. van Nostrand, Princeton, NJ

Changeux, J.-P. (1986). *Neuronal Man*, Oxford University Press, Oxford

Churchland, P. M. (1988). *Matter and Consciousness: A Contemporary Introduction to the Philosophy of Mind*, revised edition, MIT Press, Cambridge, MA

Churchland, P. S. (1981). "On the alleged backwards referral of experiences and its relevance to the mind-body problem", *Philosophy of Science* 48: 165–181

Churchland, P. S. (1986). *Neurophilosophy: Toward a Unified Science in the Mind-Brain*, MIT Press, Cambridge, MA

Churchland, P. S. & Sejnowski, T. J. (1992). *The Computational Brain*, MIT Press, Cambridge, MA

Colquhoun, D. & Hawkes, A. G. (1995). "The principles of the stochastic interpretation of ion-channel mechanisms" in *Single-Channel Recording* ed. by B. Sakmann & E. Neher, 2nd ed., Plenum Press, New York, pp. 397–482

Crick, F. (1994). *The Astonishing Hypothesis*, Simon & Schuster, London

Crick, F. & Koch, C. (1990). "Towards a neurobiological theory of consciousness", *Semin. Neurosci.* 2: 263–275

Cunnane, T. C. & Stjärne, L. (1982). "Secretion of transmitter from individual varicosities of guinea-pig and mouse vas deferens: all-or-none and extremely intermittent", *Neuroscience* 7: 2565–2576

Davies, P. C. V. & Brown, J. R. (1986). *The Ghost in the Atom*, Cambridge University Press, Cambridge

Dennett, D. C. (1991). *Consciousness Explained*, Little, Brown and Co., Boston, MA

Dodge, F. A. & Frankenhaeuser, B. (1958). "Membrane currents in isolated frog nerve fibre under voltage clamp conditions", *J. Physiol. Lond.* 143: 76–90

Donald, M. J. (1990). "Quantum theory and the brain", *Proc. R. Soc. Lond.* A 427: 43–93

Eccles, J. C. (1953). *The Neurophysiological Basis of Mind*, Oxford University Press, Oxford

Eccles, J. C. (1973). *The Understanding of the Brain*, McGraw-Hill, New York

Eccles, J. C. (1986). "Do mental events cause neural events analogously to the probability fields of quantum mechanics?", *Proc. R. Soc. Lond.* B 227: 411–428

Edelman, G. M. (1992). *Bright Air, Brilliant Fire*, Allen Lane, The Penguin Press, Harmondsworth

Érdi, P. (1993). "Neurodynamic system theory: Scope and limits", *Theor. Med.* 14: 137–152

Feigl, H. (1967). *The 'Mental' and the 'Physical'*, University of Minnesota Press, Minneapolis

Fenwick, E. M., Marty, A. & Neher, E. (1982). "A patch-clamp study of bovine chromaffin cells and of their sensitivity to acetylcholine", *J. Physiol. Lond.* 331: 577–597

Gray, C. M. & Singer, W. (1989). "Stimulus-specific neuronal oscillations in orientation columns of cat visual cortex", *Proc. Natnl. Acad. Sci. USA* 86: 1698–1702

Hesse, M. B. (1961). *Forces and Fields: The Concept of Action at a Distance in the History of Physics*, Thomas Nelson and Sons, London

Hesslow, G. (1994). "Will neuroscience explain consciousness?", *J. Theor. Biol.* 171: 29–39

Hille, B. (1992). *Ionic Channels of Excitable Membranes*, 2nd ed., Sinauer Associates, Sunderland, MA

Hodgkin, A. L. (1964). *The Conduction of the Nervous Impulse*, Liverpool University Press, Liverpool

Horn, R. (1984) "Gating of channels in nerve and muscle: A stochastic approach" in *Ion channels: Molecular and physiological aspects* ed. by W. D. Stein. Academic Press, New York, pp. 53–97

Huxley, T. H. (1893). *Method and Results: Collected Essays*, Vol. I, Macmillan, London

Jack, J. J. B., Noble, D. & Tsien, R. W. (1975). *Electric Current Flow in Excitable Cells*, Clarendon Press, Oxford

Jack, J. J. B., Redman, S. J. & Wong, K. (1981) "The components of synaptic potentials evoked in cat spinal motoneurones by impulses in single group Ia afferents", *J. Physiol. Lond.* 321: 65–96

James, W. (1879). "Are we automata?", *Mind* 4: 1–22

Jammer, M. (1957). *Concepts of Force: A Study in the Foundations of Dynamics*, Harvard University Press, Cambridge, MA

Johansson, S. & Århem, P. (1994). "Single-channel currents trigger action potentials in small cultured hippocampal neurons", *Proc. Natnl. Acad. Sci. USA* 91: 1761–1765

Korn, H. & Faber, D. S. (1987). "Regulation and significance of probabilistic release mechanisms at central synapses" in *Synaptic Function* ed. by G. M. Edelman, W. E. Gall & W. M. Cowan, John Wiley & Sons, New York, pp. 57–108

Lee, C-Y. (1992). "A possible biological role of the electron transfer between tyrosine and tryptophan", *FEBS Letters* 299: 119–123

Li, M., Unwin, N., Stauffer, K. A., Jan, Y. N. & Jan, L. Y. (1994) "Images of purified Shaker potassium channels", *Curr. Biol.* 4: 110–115

Libet, B. (1973). "Electrical stimulation of cortex in human subjects, and conscious sensory aspects" in *Handbook of Sensory Physiology*, Vol. 2, ed. by A. Iggo, Springer-Verlag, Berlin, pp. 743–790

Libet, B. (1985). "Unconscious cerebral initiative and the role of conscious will in voluntary action", *The Behavioral and Brain Sciences* 8: 529–539

Libet, B. (1996). "Conscious mind as a field", *J. Theor. Biol.* 178: 223–224

Libet, B. (1997). "Conscious mind as a force field: a reply to Lindahl & Århem", *J. Theor. Biol.* 185: 137–138

Libet, B., Wright, E. W. Jr., Feinstein, B. & Pearl, D. K. (1979). "Subjective referral of the timing for a conscious sensory experience. A functional role for the somatosensory specific projection system in man", *Brain* 102: 193–224

Liebovitch, L. S. & Toth, T. I. (1991). "A model of ion channel kinetics using deterministic chaotic rather than stochastic processes", *J. Theor. Biol.* 148: 243–267

Liljenström, H. (1997). "Cognition and the efficiency of neural processes" in this volume

Lindahl, B. I. B. (1997). "Consciousness and biological evolution", *J. Theor. Biol.* (in press)

Lindahl, B. I. B. & Århem, P. (1994). "Mind as a force field: Comments on a new interactionistic hypothesis", *J. Theor. Biol.* 171: 111–122

Lindahl, B. I. B. & Århem, P. (1996a). "The mental force field hypothesis: a reply to Libet", *J. Theor. Biol.* 178: 225–226

Lindahl, B. I. B. & Århem, P. (1996b). "The relation between the conscious mind and the brain: a reply to Beck", *J. Theor. Biol.* 181: 95–96

Llinas, R. R. (1988). "The intrinsic electrophysiological properties of mammalian neurons: insights into central nervous system function", *Science* 242: 1654–1664

Lynch, J. W. & Barry, P. H. (1989). "Action potentials initiated by single channels opening in a small neuron (rat olfactory receptor)", *Biophys. J.* 55: 755–768

MacKay, D. M. (1966). [Discussion.] in *Brain and Conscious Experience* ed. by J. C. Eccles, Springer-Verlag, Berlin, pp. 312–313

Maturana, H. R. & Varela, F. J. (1980). *Autopoiesis and Cognition*, Reidel, Dordrecht

Neher, E. & Sakmann, B. (1976). "Single-channel currents recorded from membrane of denervated frog muscle fibres", *Nature* 260: 799–802

Ochoa, J. & Torebjörk, E. (1983). "Sensations evoked by intraneural microstimulation of single mechanoreceptor units innervating the human hand", *J. Physiol., Lond.* 342: 633–654

Penrose, R. (1989). *The Emperor's New Mind: Concerning Computers, Minds, and the Laws of Physics*, Oxford University Press, Oxford

Popper, K. R. (1953a). "Language and the body-mind problem: a restatement of interactionism" in: *Proceedings of the XIth International Congress of Philosophy*, Vol. 7, North-Holland, Amsterdam, pp. 101–107

Popper, K. R. (1953b). "A note on Berkeley as precursor of Mach", *Brit. J. Phil. Sci.* 4: 26–36

Popper, K. R. (1955). "A note on the body-mind problem", *Analysis* 15: 131–135
Popper, K. R. (1956). "Three views concerning human knowledge" in *Contemporary British Philosophy* ed. by H. D. Lewis, George Allen & Unwin, London, pp. 357–388
Popper, K. R. (1972). *Objective Knowledge: An Evolutionary Approach*, Clarendon Press, Oxford
Popper, K. R. (1976) *Unended Quest: An Intellectual Autobiography*, Fontana/ Collins, London
Popper, K. (1978). "Natural selection and the emergence of mind", *Dialectica* 32: 339–355
Popper, K. R. (1984) "Critical remarks on the knowledge of lower and higher organisms, the so-called sensory motor systems" in *Sensory-Motor Integration in the Nervous System* ed. by O. Creutzfeldt, R. F. Schmidt & W. D. Willis, Springer-Verlag, Berlin, pp. 19–31
Popper, K. R. (1994). *Knowledge and the Body-Mind Problem: In Defense of Interaction*, Routledge, London
Popper, K. R. & Eccles, J. C. (1977). *The Self and Its Brain: An Argument for Interactionism*, Springer-Verlag, Berlin
Popper, K. R., Lindahl, B. I. B. & Århem, P. (1993). "A discussion of the mind-brain problem", *Theor. Med.* 14: 167–180
Puccetti, R. (1977). "Sperry on consciousness: a critical appreciation", *J. Med. Philos.* 2: 127–144
Putnam, H. (1979). *Mathematics, Matter and Method: Philosophical Papers* (2nd Edition). Cambridge University Press, Cambridge
Richards, R. J. (1987). *Darwin and the Emergence of Evolutionary Theories of Mind and Behavior*, University of Chicago Press, Chicago
Ryle, G. (1949). *The Concept of Mind*, Hutchinson, London
Searle, J. R. (1992). *The Rediscovery of the Mind*, MIT Press, Cambridge, MA
Sperry, R. W. (1977). "Forebrain commissurotomy and conscious awareness", *J. Med. Philos.* 2: 101–126
Sperry, R. W., Gazzaniga, M. S. & Bogen, J. E. (1969). "Inter-hemispheric relationships: the neocortical commisures; syndromes of hemisphere disconnection" in *Handbook of Clinical Neurology*, Vol. 4, ed. by P. J. Vinken, & G. W. Bruyn, North-Holland, Amsterdam, pp. 273–290
Stoerig, P. & Brandt, S. (1993). "The visual system and levels of perception: properties of neuromental organization", *Theor. Med.* 14: 117–135
Strassberg, A. F. & DeFelice, L. J. (1993). "Limitations of the Hodgkin-Huxley formalism: Effects of single channel kinetics on transmembrane voltage dynamics", *Neural Computation* 5: 843–855
Szentagothai, J. (1993). "Self-organization: the basic principle of neural functions", *Theor. Med.* 14: 101–116
Thompson, I. (1994). "From darkness into light", *Current Biology* 4: 458–461
Watkins, J. W. N. (1974). "The unity of Popper's thought" in *The Philosophy of Karl Popper* ed. by P. A. Schilpp, Open Court, La Salle, IL, pp. 371–412

11. Concluding Remarks

Peter Århem, Hans Liljenström, and Uno Svedin

The objective of this book was to investigate and discuss to what extent matter matters for cognitive functions and consciousness. Scientists from various disciplines, experimentalists as well as theoreticians, have here attacked the classical mind-matter problem with rather new and radical approaches. A problem such as this requires a multilevel attack from many fronts, and today this is done more than ever before. It now seems clear that consciousness actually is approachable through modern scientific analysis. It could, however, be questioned whether we are still at a pre-Copernican stage in this endeavor. Recent theoretical and technical developments in fields such as mathematics, physics, and computer science have added to the understanding of the brain previously given primarily by the more traditional fields of neuroscience and psychology, where also experimental data have increased tremendously. In this book we have been given concrete examples from biological systems and computer simulations, as well as more profound theoretical/philosophical thoughts regarding the mind-matter problem. It is not an altogether coherent view that has been presented, but rather the aim has been to highlight different facets of the problem. Some common themes have, however, kept coming back.

Perhaps the most important contribution to the mind-matter problem, is the shift of focus of interest from a traditional matter/non-matter discussion, to a discussion of what is computational and non-computational in the brain/mind processes. This shift is apparent throughout the whole book, and we believe it has brought the discussion to a more fertile ground. Another related and important point, which is often overlooked is that cognition and consciousness are not necessarily synonymous categories. When this fact has not been acknowledged it has led to confusion. Although some aspects of cognition may well be approached in computational terms, this does not necessarily mean that consciousness can be understood in such a framework.

In fact, new principles, and perhaps even a new physics, or other domains of knowledge, might be needed to explain consciousness. This view is advocated by several authors in this volume. This should be seen as a challenge for future scientific exploration. Thus, this book is colored by a strong trust that science can contribute to the unveiling of some of the mysteries around human consciousness, but maybe not within the framework of present-day theories. Our understanding is still too inadequate, but we hope this book has given some insight into the problems involved and some guidelines for possible new roads to follow.

Index

MIX
Papier aus verantwortungsvollen Quellen
Paper from responsible sources
FSC® C105338

If you have any concerns about our products,
you can contact us on
ProductSafety@springernature.com

In case Publisher is established outside the EU,
the EU authorized representative is:
Springer Nature Customer Service Center GmbH
Europaplatz 3, 69115 Heidelberg, Germany

Printed by Libri Plureos GmbH
in Hamburg, Germany